Meetings with Remarkable Kooks

Michael Theroux

Foliotroves Press

Meetings With Remarkable Kooks

ISBN 978-0-578-22126-7

Cover design by Albrecht Dürer and Michael Theroux

Published by Foliotroves Press

Printed in USA by 48HrBooks (www.48HrBooks.com)

ACKNOWLEDGEMENTS

First of all I need to thank the numerous kooks I've encountered in my life. Without them, this book would not exist. They are encountered on a regular basis to this very day.

Most of all I'd like to thank Molly Conrad for the infinite support of this project - she's endured the entire madness of my obsession with kookdom, and I'm so grateful to have her love and wealth of encouragement in the process of this work.

Many thanks for the help and support of Thomas Brown who without his guidance none of this would have ever happened. Thanks to Michael Elsey, Patrick Armstrong, Brian Butler, Gerry Vassilatos, Deva Smith, and Rob Coleman, who all in their own way supported Borderland Sciences Research Foundation.

And last to all those brilliant alternative scientists who have departed this dimension, You may well be deemed kooks, but we know who you are.

TABLE OF CONTENTS

INTRODUCTION

Some of the names presented here are factual, some not. The ones that have been changed might simply be the result of me forgetting who they are. Too many characters in this twisted tale. This is the true story of one writer's journey into the abyss of what may be known as "alternative science." It is a simple recollection of the insanity of all that I experienced over the course of 20 years as a writer and publisher in the alt-science genre. I would have liked to call this tale "Lies, Sex, and Very Little Videotape," but too much of it actually happened. Bill Pullman (as Fred Madison) said it best in David Lynch's movie, *Lost Highway*, "I like to remember things in my own way," and this just may be "how I remembered it."

Many of the people included in this treatise should not necessarily be considered kooks. They were, or are brilliant researchers whose work has been sadly misunderstood. If they are included here, it is because they fell into the realm of kookdom as I have defined it, and otherwise they should have been heralded as the geniuses that they actually were. As for my own kookiness, the mere fact that I took this on as a serious career rather than a curious avocation states the obvious.

The title of these storied words comes from a play on the title of the famous book by George Ivanovich Gurdjieff, *Meetings with Remarkable Men*. The book was written around 1927, and in it, G.I. Gurdjieff initiates an eager group of spiritual adventurers called the "Seekers of Truth," and they travel to Africa and Asia in search of the hidden mysteries of antiquity. The first actual publication date was 1963. In *Meetings with Remarkable Men,* Gurdjieff laments that people generally live their lives asleep, are not conscious of themselves, and behave like machines, subject to outside causes and imagined pressures. Also, one of the chief assessments of the work is that the people of past epochs lived in more suitable outer conditions and at higher inner levels than people do today, which I would have to say is very true of our present techno-focused society.

Meetings with Remarkable Men begins with Gurdieff's childhood, when he finds his academic book learning at odds with paranormal events that were self-evidently real but inexplicable through modern science. Later, he climbs the Himalayas, follows the Nile, and is led blindfolded to a mysterious monastery. In his encounters with dervishes, monks and fakirs, Gurdjieff ultimately reclaims the wisdom he seeks, and by comparison with what is considered Western understanding, he says, is completely backwards and barbaric.

If I had to choose a personal guru, or spiritual leader, it would have been Gurdjieff. He sold cheap rugs from the Far East to the bourgeoisie, and his penchant for forcing his vices on his loyal pupils is well documented. Initiation at its finest. Gurdjieff inspired a deep support and allegiance in his pupils - so much so that at several of his disciples' gatherings he would have them smoking cigars, slamming vodka, and chowing on beef sticks. Gurdjieff presents this as his brand of awakening - the spiritual worth of clouding up the threshold between pure allegory and unexaggerated facts. Gurdjieff very simply took his worshipful neophytes on a journey straight into the abyss from whence they came. How they emerged on the other end of the beef stick was theirs alone to craft.

The characters in this book are real, and most of them are the asleep ones Gurdjief refers to, the kooks, the machines in perfect step with the agit-prop they wholly embrace. I will apologize for none of them. To them, I can only say, "Just think about Jesus, you'll be fine watching Pat Robertson and suckling on the teat of pre-packaged microwave cheeseburgers slathered with the high-fructose corn syrup red condiment you rely on called ketchup."

It's high time we all "Catch Up."

As a child, I could have been easily labeled a kook. I had a deep fascination with all things radio, and at twelve years old, I created a pirate radio station that I named, "Revolution Radio." It was centred around a playlist of songs which at the top included the song "Revolution" by the Beatles. It was the very bumper music of the show. The transmitters and the wires that conjoined them were haphazardly strewn all over this young teen's bedroom in order to broadcast what I thought to be far more interesting than the locally offered pop AM radio transmissions. My parents were quite frightened by the entangled web of my creation, and commanded that it be dismantled lest I burn the house down.

Maybe I should have gone straight into college after high school, but I didn't go then, and have never attended a conventional college since. I hated high school. I thought the teachers were dumb, the curriculum was dumb, and most of my friends and schoolmates were dumb. I hardly ever did any homework and still managed to end up with passing grades. My teachers were disgusted with me as were my parents. How could such a smart young lad be doing so poorly in school? Well, because it was dumb. High school in North Dakota was not what you would call a mecca of academic intelligentsia. So, I educated myself by late-night readings of H.P. Lovecraft, and Edgar Allen Poe. I listened to shortwave radio at 3 am. I played chess regularly with my friend Doug (who wasn't one of the dumb ones - he was a true genius of strategy - of the hundreds of games of chess we played, I maybe won once). I studied electronics from the pages of *Popular Science*, actually read the articles in *National Geographic*, and longed for the Bugattis, Lamborghinis, and Porsches from the pages of *Car and Driver* - all of which I had subscriptions to. I learned how to play the guitar, and studied the musical compositions of Olivier Messaien, Mahavishnu John McLaughlin, and Frank Zappa. When I was in trouble for something I did, which was most often, I was grounded. I wasn't allowed to go anywhere - except the public library. There I discovered the records of the old blues cats like Son House, Robert Johnson, and Lightnin' Hopkins, and I learned their songs note for note.

"Man, you know I done enjoyed things that Kings and Queens will never have...In fact, Kings and Queens can never get. And they don't even know about it.... and good times?"

Yeah, there were good times too, but it always

centered around the mindless stuff like partying, and fooling around with girls. But, the people I went to school with had no knowledge of all these things I was about, and they didn't care as much as I didn't care to tell them.

I joined the United States Marine Corps on the delayed entry program when I was 17. One month after I graduated high school, I shipped off to Parris Island, South Carolina for boot camp. My brilliant escape. I emerged four years later in Southern California to be honorably discharged. The experience did not change how I viewed the world before I left, if anything, it enhanced the worldview I already possessed. I was no more "ship-shape" when I left the gates of Camp Pendleton than I was when I entered the gates of hell at the Marine Corps Recruit Depot. It would be many years later before I discovered that my tour in the Marines was something exceptional.

Early in 1995, I assumed the directorship of a small publishing company called, "Borderland Sciences Research Foundation." I'd been working with "Borderland Sciences" as it was known, for almost ten years before I succeeded the previous director. Borderland Sciences has a rather sordid past. It was founded in 1945 by Meade Layne as a repository and literary output of mediumship and flying saucers. Meade Layne published a magazine of sorts called "The Round Robin," a stapled-together compilation of articles from the fringes of science. Flying saucers, channeled deities, ancient occult knowledge and a smattering of alternative health remedies adorned its mimeographed pages. In 1959, another became the director of Borderland Sciences. Riley Crabb continued the flying saucer themes with an intense knowledge of the Cabbala. Riley was also an experimenter, and provided much of the material included in the pages of "The Journal of Borderland Research," the new name of the "Round Robin."

I met Tom Brown in 1985. Tom had recently become the new director of BSRF (Borderland Sciences Research Foundation), succeeding Riley Crabb. I was six months out the Marine Corps (as a radar technician) and I was working as a pizza chef at a rather fancy Italian restaurant in Oceanside, California.

THE DESERT CRAZIES

But, let's take a step back a few months. Shortly after my discharge from the Marine Corps in 1985, I moved to Joshua Tree, in the high desert of Southern California. I initially moved there only to be close to the Joshua Tree National Monument, an incredible place for climbing rocks. I had been working in Palm Springs as a security guard at GTE (General Telephone). This was in the day when female operators were still answering the calls of people dialing zero. My job was to check the badges of everyone who came through the front door. I was 22 years old, by far the youngest of all of the guards employed there. Since I had different shifts at different times, I developed "relationships" with several of the telephone operators who passed through the gates of my authority. Sasha, an aerobics instructor by morning, telephone operator by night, showed up for work early one night. She was wearing a tight, white and black geometric patterned knit dress, and was exceptionally flirty that night. She told me she was going to drive her car around to the back entrance, and that I should meet her there in 5 minutes. It seemed like a really good idea at the time. I locked the front door, and walked through the middle of the maze of General Telephone switchboards to the back door. I opened the back door, and Sasha was sitting there in her idling white Mustang GT. I looked right and left - no one was around. I opened the Mustang's door and got in. No word was said between us. After we finished our "business," I navigated back through the maze of switchboards to the front door of the GTE building, and there was Sasha, ready to “badge in” for work.

The next day I got a call from my superior at the security company I was working for. He told me that several employees at GTE had complained about me, and that I was fired. Apparently, Sasha liked to talk, and I received several nasty phone calls in the few days thereafter. I never heard from any of them again save one, who shall remain unnamed here, as she’s truly a decent human.

So now I am out of work, and living in one of the most remote little towns in all of the United States. There was no work in Joshua Tree - just flea markets, tarantulas, and drug dealers. One of those drug dealers I happened to meet at the only laundromat in town. We talked about some weird shit for awhile, and he said he could introduce me to a private investigator he knew, who was looking for ex-military guys to do some work for him. Sounded exciting, albeit fantastical, and it was right up my alley at this early stage of my spiritual development.

I greeted the private investigator, Roger Shaft, through the screen door of his suburban Yucca Valley house two days after I had been fired from the security company. After a brief exchange of small chat and casual name dropping, Roger invited me in, and he immediately began trumpeting his sleuthing accomplishments. Roger stated he was hired by a Los Angeles man to find out what happened to his daughter, and that is why he was now living in the high desert. According to Roger, a young girl had been abducted from the Joshua Tree National Monument, and her father from L.A. had hired him to investigate her disappearance.

After checking out this story, and finding it to be factual, I thought I had truly connected with my destiny. I was to become a bona fide private investigator after working with the greatest PI the world had ever seen. Roger was my new God, and was my PI messiah, and I would do all that he commanded in order to be the good student. Roger told me that he thought this desert compound he'd been keeping an eye on just might be the location of the abductors. My first task was to go check out this place called "The Integretron" in rural Landers, California. The two-acre compound (it was surrounded by a barbed-wire fence) sported two dome-shaped buildings, one large, and one small silo which supposedly housed a telescope - a mini observatory. Around this observatory circumambulated some twenty-plus abandoned vehicles - and Roger thought this very suspicious. I drove the twenty-some miles out to the remote compound at around 1:00 am, and exited my Chevy Luv truck to examine it more closely. Seconds after approaching the barbed-wire-topped chain-linked fence, frothing, barking dogs came running in my direction.

I split the scene, and 20 minutes later ended up banging on Roger's door at 2:00 am. He let me in and listened to my nutty rendition of what happened. Unfazed, Roger told me I needed to go back out there - tonight - and measure the time it took the dogs to reach me at the fence after I gave it good rattle. Reluctantly, I complied.

It took the two Dobermans almost 1 minute 50 seconds to arrive at the fence. I calculated it would be nearly impossible to jump the fence, run over to one of the cars, open the door, and grab a handful of whatever was in there before the dogs were upon me.

The next evening was the actual op. I parked about a half a mile away from the Integretron site, and walked slowly up to the fence. There were roughly ten vehicles, all in an obvious state of abandonment parked in a circle around the lonely observatory, and my mission was to commandeer something located inside one of the vehicles. I looked at my watch, threw a military wool blanket over the barbed wire, swiftly scaled the fence, and then ran toward the nearest station wagon. The dogs were in action. I reached through the back window of the wagon, grabbed a medium-sized box, and sprinted back toward the fence. I threw the box over the fence, climbed up and over it, scooped up the box, and tossed it into the back of my truck. I bet I was nearly a half mile down the gravel road before the dogs reached the fence.

It was 2:30 am when I showed up at Roger's. He was waiting for me and had the door open. I entered his house and laid the box I'd just stolen on the kitchen table. Without a word, Roger immediately started to go through its contents. I figured there would be nothing interesting in the box.

Three days before this caper was set up, Roger invited me over to his house for a "round table" discussion about this particular operation. He told me the reasons this "Integretron" compound was of interest was because of the two desert crazies that were supposedly living there, and Roger's keen observation that there were a good 30 or so "abandoned" vehicles which seemed (to him) to be geometrically arranged at either end of the place - they must belong to someone missing. The caretaker was one Don Lockwood, a well known con-man with a large rap sheet. The other resident was none other than Waldron Vorhees, and was famous as "Captain Clearlight" - a celebrity manufacturer and distributor of primo LSD. Vorhees had fled San Francisco for the High Desert of Southern California due to the DEA's pressure on his LSD distribution in the Bay Area. Roger thought, for some reason (that he wasn't telling me), that these desert freaks must have had something to do with the disappearance of the girl in the Joshua Tree Monument, and my youthful naivete bought into his hypothesis 100%.

Roger started pulling shit from the box that I had just delivered. The first thing he snatched from the box was a woman's dress - red - and covered in several, darker redish-brown stains - it looked like blood. Everything else he lifted from the box was all glass lab equipment and one plastic makeup case. Roger opened up the makeup case, and inside were several sheets of small perforated paper squares, each square roughly ½ inch by ½ inch - with some muddied stamped image on each square that was now indecipherable.

"Jackpot!" Roger screamed.

I recognized the paper squares as acid. At one time, I became quite fond of the hallucinogen after having been introduced to it in the Marine Corps. While out in the fields playing war games, I had been assigned to guard duty, and was tasked with marching around the encampment at night to make sure all was secure. I heard music coming from somewhere across the campsite. I had to investigate this. When I arrived on the tent where this music was coming from, here was one man and his guitar playing songs I'd never heard before. Turns out, he wrote them. Sgt. Larry Fleck introduced himself and continued playing, not missing a beat. I stopped to listen....it was brilliant. He handed me the guitar as if knowing I knew what to do with it. I did. I played him "Mood for a Day" by Steve Howe from the Yes album, "Fragile," and we were immediate friends. Larry then asked me if I wanted to "turn on my shit." Always happy to be a part of the game I had no understanding of, I said, "Sure" and he handed me a small square of paper imprinted with a picture of Mickey Mouse.

"Just put it on your tongue"

"Why?"

"Just do it, Jarhead."

It was an order from a superior non-commissioned officer, so I had to comply. The rest of my stay in the Marine Corps was never the same.

Sometime later, Larry and I formed a band called "Larry Fleck and the Flecktones." Our band played many of the venues on the Camp Pendelton Marine Corps base, and we were equally kicked out of all of them for being either too loud, or just awful. Unfortunately, Larry was killed in a car accident shortly after his discharge from the Corps. I'll never forget Larry and what he did for me that innocent night I was on patrol.

Roger slapped me on the back and said, "Good job." Four hours later he was on the the phone with the State police. He told them of the evidence he had, and they completely blew him off. I later found out they thought he was a total kook and dismissed every single thing he had prodigiously presented them.

My entire stint working with this private investigator lasted only a few weeks before I realized he was nothing but a confidence trickster and degenerate clownfestival - and I was an idiot.

A great friend I'd been spending all of my other time with offered a grand solution for me. She knew I wanted out of the High Desert scene, and suggested her 1968 VW Westfalia microbus would be a good trade for my little pickup truck - I'd always have a place to stay if I had a camper van. We traded our precious vehicles one for one, and I was outta that fucking nightmare that night, I drove straight through to North County San Diego, and camped out on South Carlsbad State Beach. The next day I called a couple of old friends from the Marine Corps and was able to secure a room-for-rent in a friend's house in Vista, California - just inland from Carlsbad. I also got a job that very day at an Italian restaurant in Oceanside slinging dough. Things were looking better.

About a month after I'd moved in with the old jarhead pal, I got a call from Roger. I almost hung up on him, but I was too curious as to why he had tracked me down, and how he did it. He never really told me, but he did want me to do some work for him. Roger asked me if I remembered the Integretron. How could I forget. He then stated that he had on some desert crazy authority that a "cult" group called "Borderland Researchers" was somehow affiliated with the Landers compound, and that they were located right here in Vista. Roger tasked me with contacting the "leader" of Borderland Researchers (according to Roger, the leader's name was one "Eric

Dollar") and to see if I could find out what they were really doing at the Integretron. Roger said he would pay me for any intel I could send him on this, and it seemed like an easy task, so I accepted the job.

"Borderland Researchers," I soon found out was "Borderland Sciences Research Foundation." Trouble was, they only had a P.O. box address - "P.O. Box 549, Vista, CA." But, there was also a phone number. So I called it. A man answered the phone.

"Hello, Borderland Sciences?"

"Hi......ummmm....hey, how are you?"

"Fine, what can I do for you today?"

My first real social engineering encounter had begun.

"Well, I'm a freelance journalist doing a story for the Hi-Desert Star in Yucca Valley, and I wondered if you would be open to an interview about your involvement with the Integratron in Landers?"

"Sure, no problem."

As we were setting up how to meet and do the interview, I found out Borderland Sciences was a company that was being worked out of a house - only two blocks from where I now lived. The next morning I walked the two blocks to meet a man named, "Tom Brown." He greeted me at the front door of a very small single story house. Tom told me he had recently taken over the directorship of Borderland Sciences, and gave me a tour of the operation. Borderland Sciences Research Foundation was a small publishing company - the library of old books he first introduced me to - housed in the office area - was most impressive. In the postage-stamp-sized backyard were two metal sheds - one containing previously printed staple-bound books - the other, harboring an old mimeograph machine oozing blue ink.

This odd publishing company completely fascinated me. In the 1970s, I was an avid fan of a TV show called, “In Search Of” with one of my great heroes, Leonard Nimoy or “Mr. Spock” of the original Star Trek series narrating. One episode of this show particularly captivated my sense of wonderment. It was a documentary about Coral Castle located in Rockgate, Florida, and the lanky, 120-pound immigrant that built it for his “Sweet Sixteen.” Coral Castle was an enigma of architectural genius - 20 and 30-ton blocks of coral carved with precision, and adroitly arranged into a fortification complete with massive revolving doors you could open with a finger, and otherworldly megalithic pronouncements of squarely-placed tonnages. It was entirely crafted in secret by one little man named, Edward Leedskalnin, from the cutting of the blocks of coral from a quarry many miles away, to the exacting erection of every stone. During the show, Nimoy talked briefly about a book that Edward had written entitled, “Magnetic Current” with a picture of the cover displayed. Supposedly, as the story goes, the secrets of the building of the pyramids in Egypt and the construction of Stonehenge were revealed in this book. Naturally, this put me on a quest to find the little book of

Leedskalnin's. No bookstore had ever heard of it, and the Library of Congress had no record of its publication. Seemed to be a dead end.

Tom Brown and I hit it off quite well that first encounter, and he called me a couple days later asking if I wanted a job in the office. The first order of the job was to sort through the massive library of books on the numerous eight-foot high shelves, and arrange them by categories. I jumped at this opportunity and started in on it the next day. Here's where the story clearly defines my own kookiness.

It just happened to be my 23rd birthday, and here I was eight feet up on a ladder in the Borderland Sciences library, pulling dusty tomes from the top shelf of the homemade bookcase embedded into the wall. The trouble with this endeavor was that nearly every book I pulled and opened caught my fascination, and I think I did more reading than arranging while standing on that ladder.

While pulling another book from the shelf, a small pamphlet fell cover up on the floor below. It was "Magnetic Current." I nearly fell off the ladder trying to get to it. Sure as shit, this was my grail book in the flesh, and the work day was over. I took the 19-page secret-of-the-universe book home with me, and then read it cover-to-cover....5 or 6 times into the wee hours of the morning. I didn't understand a fucking word of it. The book has no diagrams and talks of south and north-pole magnets and how to do little experiments to prove his hypothesis. No secrets of the universe whatsoever.

The experience did do one thing for me. It made me believe that miracles were indeed possible, albeit the outcome of them might not be what you expect them to be. While my fantasy to be the discoverer of Edward Leedskalnin's secret on how he constructed Coral Castle was crushed, it only bolstered my curiosity to find other secrets hidden within rare, lost manuscripts. I was hooked, and kept working at Borderland Sciences for the next 14 years. Later in those years I would discover that Ed Leedskalnin probably made up the entire mystery of the secret in that book. His workshop was found to be filled with come-a-longs and pulleys and ropes - all of which he would have never needed had he really known the mystical riddles of the ancients' megalithic construction techniques.

A few weeks into my new job I told Tom about the private eye I had been working for, and his theories about the Integretron.

"Who is Eric Dollar?" I asked.

Tom laughed and said, "You must mean Eric Dollard. He's an electrical engineer that's been rewiring the Integretron to restore it to its original purpose. A time machine of sorts....errr, rather a giant rejuvenation device."

Tom went on to explain the history of the Integretron. The Integretron is a rather large, dome-shaped building that was originally constructed by a man named George Van Tassel. It is located on a lot near a place known as "Giant Rock." Giant Rock is a single free-standing boulder roughly the size of a two-story house, and was considered a sacred site by native americans. In the early 1930s, a man named Frank Critzer staked some mining claims in the Giant Rock area. He soon thereafter built a small one-room subterranean home underneath the boulder. Critzer became well-known as a true desert crazy and was avoided by most of the local residents. Van Tassel met Critzer some years later and was invited to visit Giant Rock. During WWII, Critzer came under suspicion as a German spy, and was killed in a government raid on his dwelling in 1942. Supposedly, all of Critzer's possessions were removed by whatever agency had raided the stone. The room he had built underneath the rock was burned out, possibly by an incendiary device during the raid, and for many years it was locked up and closed. In 1947, George Van Tassel, after having been granted a lease to the property from the Bureau of Land Management, packed up his belongings and family, and moved to the Giant Rock Site.

Critzer had already established an airfield at Giant Rock, and Van Tassel, a flight test engineer for both Lockheed International and Douglas Aircraft, reopened the runway complete with a new cafe. Van Tassel had spent the World War II years flying with Howard Hughes. There are stories that Howard Hughes would fly into the Giant Rock airport on weekends just so he could get a slice of the pie that Van Tassel's wife made. But, it has been suggested that the real reason for Hughes' visits was that he knew of Van Tassel's plans to build a fountain of youth device, and Hughes was known to be a seeker of eternal life (he wanted to be cryogenically frozen until research could find a cure for whatever would be his demise).

Weekly fly-ins were not uncommon, and Van Tassel began having UFO conventions at Giant Rock in the early 1950s (it was believed by many UFOlogists that Giant Rock was placed in that location by flying saucers). In 1959, over 11,000 people attended his Giant Rock UFO convention.

Van Tassel conducted many mediation sessions under Giant Rock, and believed he was channeling alien entities who told him how to build a rejuvenation machine. He created the building with this direct input from his alien designers, and in doing so generated a huge following of others who believed he had tapped into real alien knowledge. Van Tassel believed he had received instructions for a machine that could rejuvenate human cells using the natural energy found in the atmosphere. He called this apparatus/dome the "Integratron."

Van Tassel funded the Integratron construction by using Giant Rock as a focal point for his flying saucer conventions. The first convention was held there in 1954, and thousands attended these events - from the stories told, and the many photographs taken at these events, it was not unlike the modern-day Burning Man festival, and may just be the forerunner of these desert crazy parties.

Unfortunately, Van Tassel died before the Integretron was finished, and finally, in late 1985, the researchers of Borderland Sciences were there to complete the job - this time without any alien involvement.

I dropped some more names on Tom that Roger had given me.

“Do you know who Don Lockwood and Waldron Vorhees are?”

“Hehe, of course. Don is the old bastard caretaker of

the Integretron, and Walt just hangs out there and works on cars. He's helped us with the rewiring of the dome," Tom stated.

So far the story is certainly quite kooky, but doesn't seem to warrant the attention of a PI investigating a missing child. Tom thought it might be fun to call this Roger Shaft, PI and hear his nutty theories, first hand. I got on the old rotary dial phone at Borderlands, and called Roger - Tom would be listening in on the other wireless phone.

"Hey Roger."

"Michael! What the shit is happening, man?" Roger seemed excited.

"Well, I'm calling because I need a little more info on this whole Integretron thing. I tracked down the Borderland Researchers..."

"Outstanding! Let me tell you this fucking thing is getting just crazy. The State Police are all over it, and I just did two radio interviews in LA this weekend."

"Wow, Roger, that's cool. So fill me in again on this thing...what's going on out there?"

"Ok, so the dome (Integretron) we figure is being run by this guy named Eric Dollar who is the ringleader. He's the cult guru, and Don and Walt are his henchmen. They seem to be running some kind of thing where they abduct people from around the area, make them disappear, and steal all their money and belongings - it looks like that dome building is where they're storing the stuff - I bet it's just lined with cash."

"Really? How can you be sure?" I asked.

"It's obvious because of all the abandoned cars on the lot, shit-for-brains. So whataya got on this Borderland Researchers outfit?"

At this point Tom answers.

"Hey asshole, how can you possibly call yourself a private investigator...you couldn't private investigate yourself out of a paper sack?"

".....Who the fuck is this?" Roger snapped back.

"I'm the director of Borderland Sciences, idiot. You have no idea what we're doing out there, and I suggest you stay the fuck away from us, or someone may just find your abandoned car rusting away on the lot."

Roger hung up. I never heard from him again. He certainly never solved the missing girl case, and disappeared into total obscurity.

There seems to be something about deserts that makes people completely nuts. I don't think it is the daily heat, the nightly cold, nor the dry parchedness of the environment. I think it is due to the raw isolation inflicted upon its willing denziens. The people of the world's deserts possess an undefined kookiness that borders on absolute insanity, the likes of which is not observed anywhere else on the planet. I experienced it in others, and in myself while living there. “A fun place to visit, but never live here,” was intoned by many who had made the desert their home. I left there after a stay of one single year, and I never looked back. Ok, I did visit a few times in later years - in each case, it became (due to uncomfortably creepy experiences), a place I wanted no further part of in my life. More on this later.

HOLLYWOOD EXCURSIONS AND VAPID DIVERSIONS

In the early 1980's myself and a couple of friends used to go to Hollywood. Every Weekend. We were Marines and we were stationed in the High Desert at 29 Palms - the Marine Corps Air Ground Combat Center (MCAGCC), about 2 ½ hours from Hollywood. Staying on the base on the weekends was full of the usual horrors enlisted men faced - go to the enlisted club where a bunch of other Marines (male ones) are getting drunk, so you get drunk, and all that's left to do is to fight. One of these Marine friends of mine had the brilliant idea that we go to Hollywood instead of hanging out on base.

Now, we'd explored the town of 29 Palms, "The Stumps" they called it, as someone in the town thought it a good idea to chop off the palm trees lining main street about 5 feet from the ground - more desert crazies. Then we'd gone all the way down the hill on 29 Palms Highway through Morongo Valley and ended up in Palm Springs. We were simply looking to get laid. The trouble with that, at the time, was that long hair was in fashion. As Marines, we had none of that. So most places like clubs in Palm Springs were never going to work for the intended outcome we were seeking. My brilliant friend Ron realized there was a punk-rock scene happening in Hollywood, and the fact we had no hair was perfect. So, we went to Hollywood, and we fit right in. Sort of.

There's a good reason this chapter is in the book. The book is about kooks, and Hollywood harbors more kooks *per capita* than anywhere on the planet. That's why it's affectionately known as, "Hollyweird."

Everyone's first trip to Hollywood consists of taking the long walk down the "boulevard," crossing over all the stars in the sidewalk with names embedded in brass known and unknown to us. In 1981, the heart of Hollywood was the crossroads of Hollywood Boulevard and Vine St., better known as "Hollywood and Vine." On the edge of these crossroads was the Pussycat Theatre, a porno movie house opened 1974, with nightly features and sticky floors. Two doors down was Gilbert's Book Shop, an aging fixture of Hollywood decadence where at various times in history Charlie Chaplin, Bette Davis, Greta Garbo, Joan Crawford, and Bing Crosby all used to frequent. My pal Ron had Gilbert's listed as a place we'd have to stop "on accounta" they had an enormous occult book library, and both of us at the time were interested in the works of the great magician nutcase, Aleister Crowley. We had been in search of a specific treatise though - the fantastical though highly fictional grimoire known as the *Necronomicon* or "Book of Dead Names" created by, and first appeared in the horror works of H.P. Lovecraft - specifically in Lovecraft's 1924 short story "The Hound." Lovecraft claimed the title of the book came to him in a dream. But, the version of the book we were after was, according to its new discoverers, real, and supposedly written in Sumeria in the 8th century A.D. by the "Mad Arab" Abdul Alhazred. We had read (in a comic book classified) that the proprietors of Gilbert's Book Shop in Hollywood had original, gilted, leatherbound editions of this book for sale, so we went to check it out.

In the late 1970s, a book purporting to be a translation of "the real" *Necronomicon* was published. This book, introduced by a man named only "Simon," had very little connection to the fictional Lovecraft Mythos but instead was based on Sumerian mythology. It was later dubbed the "Simon *Necronomicon*". The book went into trade paperback in 1980, has never been out of print since, and has sold over 800,000 copies.

The book was featured as courtroom evidence in the murder trials of Roderick Ferrell and Glen Mason, with suggestions that it played a part in Satanic human sacrifices. Ferrell, it is claimed, used the book during some kind of vampiric occult rituals. Whatever the case, we hadn't heard about any of this. We thought the book fascinating from an historical occult point-of-view, and our naivete and youthful enthusiasm drew us to seek out the sordid manuscript.

The old gentlemen proprietors of the shop greeted us upon entering. We informed them we were in search of the "authentic" first edition of the Necronomicon. Both of them chuckled in unison,

"Ohhhh, yes, we have it. In fact, hehehe, we have two copies in stock."

One copy was bound in leather, gilded in gold, and carried a price tag of $200. The other, cloth-bound and gilded in silver was priced at $100. Ron paid the $200 for the leatherbound edition, and I happily paid the $100 for the cloth-bound edition. We left the store quite happy that we had completed our mission - each of us cradling an expensive volume wrapped in layers of tissue paper.

The USO Club on North Ivar near Hollywood and Vine, has a rich history. It was the first facility of its kind in the nation, and became a model for the chain of USO clubs chartered by Congress in 1941 to serve the nation's military personnel overseas. Myself, Dave, and Ron thought we'd check it out on one weekend excursion, and it wasn't really any different than we expected. That day we'd wandered the mile or so of action of Hollywood Boulevard and basically just shopped. There were all manner of shops - bookstores with nothing but rows and rows of *Dianetics* by L. Ron Hubbard on the shelves, B&D shops with handcuffs and cat-o-nine tails prominently featured, *Frederick's of Hollywood*, and the Hollywood Magic Shop. We walked in the magic store and there was a large man doing tricks for a group of about 10 children. It was Muhammad Ali. He was terrible at magic, but the kids didn't care, and neither did we. As a young lad having watched all three fights on live TV with Joe Frazier, and having seen the "Rumble in the Jungle" fight with George Foreman, Ali was one of my childhood heroes. We shook his hand moved on down the boulevard to the USO club.

This "dance" club, called the USO was a complete throwback to the 1940s and WWII. There were rows of hired young women lined up around the periphery of the dance floor waiting to take the hand of any serviceman that would ask politely for a dance. I thought it was just plain creepy, but my inquisitive nature led me to ask one of the women to dance - so I could pick her brain on why she was doing this stupid fucking job. I scouted around the perimeter for someone who was different, someone who looked out of place. Dave and Ron were already tripping it up with a couple of girls - I watched them for a few seconds, and their interaction and movements looked awkward and synthetic. Then I found her. She was alone and looked tough and distant - her presentation was wild, radical, subversive and dangerous. She was my Marla Singer from the movie "Fight Club" in a perfectly dirty dress looking for something more in the patterns that the tiles on the floor might reveal to her. She was the one. I walked over to her and held out my hand. She kept looking down at the floor like she didn't even see me, so I took her hand and gently led her to the dance floor. We danced for what was left of the slow song - a song I don't remember - but one that should have evoked an embrace rather than the crude distance that occurred between us. I never said a single word to her, and the song ended. She disappeared back into the periphery, and it was time to get the fuck out of the twilight zone.

Dave and Ron were as ready to leave the USO Club as I was, and we exited into the parking lot. Shit, the parking lot was huge party festival compared to what was going on inside. We hung out for awhile and finally decided it was time to get back to the hotel. We had a room at the Holiday Inn just across from the Capitol Records building on Vine St. We could have easily walked the short distance, but we were drunk and 3 large bikers we were partying within the USO lot asked us if we needed a ride. We had no idea who they were or why they hanging out in the parking lot of the USO. I offered them my brilliant idea that we could all go back to our room at the Holiday Inn and continue partying. Things then got ugly. The bikers accused us of being "fags" and one of them flipped open a giant switchblade and commenced to wave it around in our faces.

Without hesitation, the three tough Marines ran like hell from the biker with knife and we made it safely back to our guarded Holiday Inn haven.

Over the next year while we were still stationed at 29 Palms, we made several more trips to Hollywood. Every instance was a new and predominantly dissonant adventure - we met and got into a heated argument with David Lee Roth at a stoplight, we ran into an obnoxious Shelly Winters at a diner on Melrose, bought homemade cassette tapes from a then unknown leather band named Motley Crue, who had set up a merchandise card table on the boulevard, and we partied with Black Flag and Johnny Lydon - and didn't really know who the fuck they were. Hollywood was a catalyst for embracing the divine kook - they were everywhere, in all of their sordid manifestations, famous and not, and they were a ubiquitous necessity which would prove instrumental in shaping my future relationships with the kook world.

BORDERLAND SCIENCES

Working at Borderlands was great. I eventually quit my part-time pizza-twirling job and did my 40 hours or so a week at Borderlands. I traded a 30-minute commute to a restaurant job for a two-block walk in the morning in SoCal fog to attend a feast of rotting books on dusty shelves. I started learning about shit hardly anyone one knew: Ruth Drown and Radionics, Trevor James Constable and Weather Engineering, Nikola Tesla, George Starr White, T. Galen Hieronymus, Albert Abrams, The Philadelphia Experiment, The Bermuda Triangle, and The Secret Life of Plants. I found my calling early in life, and it was grand.

About three weeks into this new intellectual fling, Tom called me into his office.

"We're moving to Northern California - in the woods. You wanna come with us?"

"Uh, okay." I said, a little unsure of myself.

"I have to ask you about something. When we met on the phone for the first time, you said you were doing a story for the Hi Desert Star in Yucca Valley. Are you still doing that?"

"Shit. You got me there....that was a total ruse, man. I was still working for that PI and that was my line."

Tom laughed, "Good. I thought it was total bullshit. So we're moving in two weeks. If you want, you can even bring that cute little redheaded girlfriend of yours."

I had been seeing this waitress from the restaurant named Liza, and I'd brought her by the new digs the week before. Liza was enamoured with my science geekiness, and seemed fond of my new job and the house full of books. But, I was really in love with another girl. I had known Kimberly for nearly two years, and while we had been together briefly in the past, now we were really just good friends. So, rather than ask Liza to move to Northern California with me, I asked Kim. Kim politely turned me down - she was already moving to Rhode Island to hop on her friend's sailboat, and gunkhole down the Eastern coast to the Virgin Islands.

It turns out that no women moved with me Northern California, and I'd say that was a good thing.

I packed up what little belongings I had, left the VW microbus parked on some side road with the keys in it, and jumped aboard the 27-foot U-Haul headed for Humboldt County. I found out from Tom while we were heading north on I5 that we would make a stop in Los Angeles to attend the U.S. Psychotronics Association annual meeting.

A few weeks before I found out I was moving, and after having read several of the libraries' books while I was perched upon the ladder, I decided to construct one of the most famous of all kooky devices. This device was the Tesla Coil. The basic Tesla Coil is an electrical device that when powered up, will emit electrical discharges that emulate lightning. Mind you, this is not what Tesla had envisioned when he created this device near the turn of the 20th century - it was not to be a spark generator - it was a technology that was to herald in the new age of energy distribution to be utilised over the entire planet. My version was a simple electrical discharge generator. Tom saw my completed device in operation, and made sure that it was carefully included in the list of my belongings - we were going to demonstrate it at this Psychotronics convention.

Psychotronics is a term coined in 1967 by Zdeněk Rejdák for the study of parapsychology using electronics and radio technology. The conference would be loaded with technogeeks of all walks - most of them quite well-versed in all things electronica. We arrived at the conference location mid-day, unloaded my crudely-built Tesla Coil, and checked in with the receptionist. The conference room was small - about 50 people were in attendance. We sat on metal folding chairs and listened to numerous presentations expounding on the metaphysical this or that of each contributor's worldview. My Tesla Coil was the only physical prop at the entire psychotronic circus, and was saved for last. I was to present my apparatus at the very end of the conference - and I did, but I had no prepared dialogue to accompany it. It powered up, and six to eight-inch blue discharges erupted from the small brass doorknob affixed to the top of its secondary coil. Ooos and ahhhs percolated from the audience, and several of them asked if they could buy it. The sale of that cobbled-together lightning generator paid for the gas for the rest of the trip north.

LSD, AND THE NORTH WOODS KOOKS

We arrived around 3pm at our new home in Whitethorn, California nearly 15 hours after leaving the Psychotronics kookfestival in LA. At the head of the winding gravel driveway, we were met by a man I only remember as John - shoulder-length hair and a Billy Gibbons beard covering any real expression on his face. John was a neighbor who was very eager to help us get settled in. We followed his mini-truck through the tallest trees I had ever seen, and after about a mile down the twisted unpaved road, we reached a house and a large outbuilding. The new headquarters of Borderland Sciences Research Foundation. The publishing operation was to be housed in the barn, and the house would be occupied by Tom and his wife. A half-mile up the road would be where I lived - a single-wide trailer on stilts that appeared to be at least 40 years old. Tom, John and I unloaded the entire U-Haul truck into the barn in about 2 hours. The few things I had would make their way down to the trailer. John had been sucking down some cheap domestic beers the entire time we were unloading, Tom and I were ready for a couple of those, and we all started drinking. At some point, John said he had some acid and would we like to drop with him. Tom and I declined, while John popped a tab in his mouth and then repeatedly stuck out his tongue at us with the square of LSD resting on the end of it. Soon thereafter, Tom retired, and I was left with a tripping John a half a mile away at my new dwelling.

The trailer was narrow and appeared unfinished on the interior save for an old Franklin stove, and an old two-burner tabletop hotplate. It looked like something a hobo would have loved. John and I lit some smokes and sucked down a few more beers. Attached to the trailer was a length-long homemade redwood deck, and John and I leaned against the railing, taking turns telling tall stories. While our stories innocently started out with girls, and drinking, and other generalities two new male acquaintances are bound by law to make up, John's stories started to take a turn in the direction of the Bible. I was not a fan of Christianity, and I finally told him as much, and then went inside to fetch another beer. Just as I was coming back out of the door of the single-entrance trailer, something whizzed past me and stuck in the faux-wood panelling of the wall next to the stove. I fell backwards, looked up at the wall, and there was a 2-foot piece of rusted rebar sticking out of it.

"Satan be gone!" John shouted.

John then attacked me in the entrance of the trailer with another piece of rebar. For some reason, the gravelled front of the place was littered with two-foot sections of iron rebar.

"I am Jesus, the Lord God, and you are an infidel. You are going to hell tonight."

What the fuck? It's nearly 3 am, and I'd been driving all day long after having endured an extremely boring three-hour detour in LA, to move to a place I had no idea even existed, and I was about to be killed by some Jethro Tull-looking asshole who thought he was Jesus?

Chuck Palahniuk couldn't have conjured up this goofy fight scene.

We struggled on the floor of the trailer for a few minutes and I was able to get away. I ran the half mile down to the house that Tom was living in and banged on the door. Tom reluctantly answered, and I told him what was going on. Tom came out onto the drive to see what was was happening, and there was nutty Jesus, both hands swinging his holy rusted rebar weapons of the Lord. Tom punched him in the face, and eventually, the local sheriff was called, and our friendly neighbor cum shapeshifting messiah was hauled away in handcuffs. My first night in the North woods of Humboldt County came to a close, and all was as quiet. Quiet as it should be. It was 4am.

GETTING FIRED THE FIRST TIME

Getting fired is an horrible experience no matter what the circumstances of your being fired were. Even if it was completely your fault, you will still hate the boss that fired you. You will make up any excuse - the boss is a dick, they have no idea what they're losing by firing me, I'm irreplaceable, and, the boss is a dick.

I was sitting on the deck attached to my trailer dwelling, fiddling with various electrical components to make a new, improved Tesla Coil, and I see Tom walking up the gravel road toward me. Tom's got a t-shirt on that says, "GET ON THE BUS" with a silk-screened thumb pointing the way to the door. I never saw that shirt before this day, and had no idea what it meant.

"Get your shit together, and get the fuck out!"

"Whaaaa?" I sort of said.

"You're done here. Pack up your shit and leave. I don't care where you go. Here's a check for $250. Just leave." Tom was obviously not happy.

"What did I do?" I knew what I did, but I had justified my actions in my own mind.

"I'm not going to explain it to you. You know what you did....hurry up."

So here's the deal. A few weeks before, Tom introduced me to a neighbor - an old-timer - who had an enormous collection of violins. The old man led me into his den, and on the wall were maybe 30 instruments, pristinely hanging there like you might find in some orchestral museum. I was not a violin player, but I was an aficionado of the arts lutherie, and marvelled at the assemblage before me. The old-timer felt my enthusiasm for the collection, and said, "Pick one one of them fiddles out....it's yours." A very generous gesture. After looking at all of them for some time, I chose a Maggini style violin with its characteristic double purfling. It was a gem of an instrument, and I knew it. What I didn't know was that you don't take such a gift so lightly. A week after this, Tom took off on a trip to the desert. I was left in charge of the publishing operation. The closest music store was up in Arcata, about an hour from where we were at. I got up one morning a few days after Tom left, and drove up to Wildwood Music with the violin in tow. When I opened the case for the owner of the store, his eyes bugged and he asked what I wanted for it. I said I had no idea, but I might be willing to take some stuff in trade for it. I walked out of the music store with a Celtic harp, an acoustic guitar, and a host of accessories in trade for that brilliant violin. Since I wasn't a violin player, I thought this was a good deal for me. When Tom arrived back from his research vacation, I showed him what a glorious trade I had made at the music store. He said nothing. The next morning he was standing on the deck in front of me, justifiably firing my stupid ass.

At the time, he was just a dick boss who had no idea what he was losing by getting rid of me. I didn't understand what I'd done at the time. Essentially, I had betrayed his trust in me, and it wasn't long before I came to understand what a shitty thing I did.

I packed up my belongings, got on a Greyhound bus, and went back to my hometown in North Dakota. It was a rough lesson, but it was a lesson desperately needed. Back in the homeland, I settled into a job at a local music store. I had told myself that I would never return to the place that I abhorred as a teen, but there I was. And, in later years, the irony of the event would not be lost on me as I became the manager of the orchestral department in another music store. But, three years after I was fired from Borderland Sciences, Tom called to offer me my job back - as managing editor of *The Journal of Borderland Research.* I didn't hesitate to accept, and I was soon packing up my newly acquired 1974 Chrysler Town and Country station wagon (I traded an old Fender Stratocaster for the car), and I headed back to Northern California.

BORDERLANDS 2.0

After a long, and very strange trip, I arrived back at Borderland Sciences Headquarters in Whitethorn, California ready to start the restart of my original job as managing editor. While on hiatus back in North Dakota, I'd penned several articles on various kooky topics ranging from weather engineering to the luminiferous ether, and even some shit about chaos theory that were all published by Tom in the *Journal of Borderland Research.* I was 28 years old and believed that I was a rather prestigious internationally published author. The "Journal" had a subscribership of around 5000 readers in 32 countries, so, to me, this was mostly true.

What I hadn't told Tom when we worked out the details of my return, was that I had gotten married and that I was bringing my new wife and two children with me. Oh yeah, I failed to mention that I was bringing an entire family. Since there was now someone living in the old single-wide trailer I'd previously lived in, and the house was occupied by Tom and his family, Tom improvised and said we'd all have to live in this "other" trailer located who-knows-where on the property. Tom lead us down the main gravel road to our new digs. Nestled back in a turn-out on the side of the road, and covered with black mold and fallen branches, was a 12-foot long travel trailer - like you'd see being pulled behind a filthy F-150 on a weekend deer hunting outing. This was to be our new home. My wife, of course, was not impressed. The children, on the other hand, were elated. "Are we camping, Daddy?" "Yes, we're camping...and we may be some time." I used a line from an old story I'd heard about one of Shackleton's South Pole expedition crew saying the very same thing to another member of the failed party, as he left his tent and never returned. If only I had the balls to be that guy right now.

It wasn't long after this that the wife packed up the car with all her belongings including the kids and left me. Who could blame her. What a shithole. But, she did move back to North Dakota, which I considered to be the biggest shithole of all.

After this, I became rather depressed, but I completely understood her decision, and the new freedom I had allowed me to immerse myself in my work, and work I did. When I wasn't editing the nutty ramblings of some deluded free energy researcher for the Journal, I'd be banging out some new esoterica I'd dreamed up on an old Smith-Corona manual typewriter. I procured the antique at a pawn shop for $15 in some unmemorable small town I'd driven through in western Montana on my way out to California. Several of the keys were sticky, mostly the ones that hardly anyone ever used when writing letters. I used that machine to write a number of articles, and an entire book on music. It never failed to assist me when I needed to write. This mechanical toolbox became especially useful when I decided to move out of the hunting trailer to an off-the-grid cabin on a mountaintop some miles from my woodland workplace. The cabin had no power and no running water. The only way to get clean was from a crappy solar shower, or the attached bathhouse whose bathtub needed to be filled with kettles of water, heated from the nearby Franklin stove. A single 3 x 1 foot solar cell charged a 12-volt car battery during the day, and this happened by whatever sunshine deemed fit to show it some attention. Usually, I had enough light in the evening (provided by 2 DC lightbulbs) to crank out short articles and ad copy for the Journal.

I became somewhat of a hermit, enjoying nearly nothing, save the reading of strange arcane manuscripts, and the attendant epiphanies which I believed resulted. I shat in a hole of my own digging, rarely bathed, and lived on leafy greens and garlic. I was my maker incarnate, the embodiment of all the ancient Gods before me...I was pure gnosis. And I needed a fucking bath and shave.

When you are truly alone, you will either embrace the inevitable onset of insanity, or you will find ways to entertain yourself in a way that you are able avoid insanity. Thank you, manual Smith-Corona typewriter with your frustrating sticky digits, and noisy response to my words, for providing me with enough support to accept your inanimate friendship as soothing company when I had nothing else of emotional warmth to embrace.

I eventually tired of the caveman life and was able to move into the house next to the barn where we published the Journal. Tom had moved to another house several miles away to get away from being so close to work, but that was just fine for me - the house had heat, running water, and real bathrooms with real showers. The running water was actually pumped up from the Mattole River upon whose banks the house sat. I hadn't realized that this pumped-in water had no filtration system on it, but it hardly mattered at the time.

A few weeks after I was fully nested in the new house, a weird smell began to permeate the area. This wasn't the same smell I had recently encountered while trying to heat up the gas oven for a frozen pizza. That smell, I unfortunately figured out, was from a certain mouse and her little nursery of mice-lings that had made their home in the broiler of the oven. I was quite horrified when I pulled open the drawer of the broiler and discovered that scene, but I quickly scooped up the smoldering mass of broasted rodentloaf with a metal spatula, and deposited it in the grass outside. The pepperoni pizza I'd impatiently cooked on the upper level of the oven still tasted great, even though it had a slight hint of mousy.

This new smell that permeated the area was hideous. A few days after one detects such an insidious olfactory assault, they will generally become immune to it - unless its source is one of decomposing flesh. A few days later, while doing dishes in front of the big window that directly looked out over the river, I spied a large number of law enforcement officers that were gathered about directly across the river, accompanied by a full team of divers and dredging equipment. It turned out that the neighbor across the stream, just upriver from me, had been sunken into the bedrock after having been murdered, and I was apparently drinking the unfiltered remains of his bloated carcass...

So, now that I was clean, fed, and warm, I found myself unable to write. But, there was still the editing, design and layout, and most importantly, the reading of submissions for possible inclusion in the Journal. Some 90 percent of these submissions, ranging from theories on how to heal one's maladies with paper cards imprinted with geometric diagrams, to how barcodes on food labels were poisoning us, were just too far out at the time for me to consider. And, many other article submissions were tossed out after reading just the first paragraph - they were unintelligible scribblings - and would have been easier to understand had I randomly anagrammed every word. In many cases, submissions would be littered with mathematical formulas and very little or no accompanying explanation. They were tossed into the bin as well. What we were really looking for were articles from researchers who had actually created a device or method that could be tested, and had been tested by the authors. Only about 2 percent of all of the submissions we received fell into this category. There were ground antenna systems designed to receive radio through the earth, historical essays on Victorian scientists complete with how-tos to recreate their inventions, articles on Tesla, Rudolf Steiner, and Dinshah Ghadiali. But there just wasn't enough good material to fill the 48 pages of the magazine. We were even so critical of the ads we received that we rejected a good many of them, and this did little to help the income of our small publishing company. So we began to write much of the material ourselves, using our real names back then. We even created fake ads, or ads that were for things we were selling. It worked out quite well.

Not long after things seemed to be rolling along nicely, I was fired again - the reason was not so important as the outcome. Yes, I would again be hired back, but in the months after this second firing, I held many jobs in the local area including: exotic animal farm helper, editor and typesetter of a Hindu religious magazine, caretaker of a local music studio, and last but not least, guarding greenhouses filled with marijuana plants. Obviously, the last job proved far more lucrative than the others, but was by no means more fulfilling. In fact, it was very short-lived, and I was soon asked to come back to the helm of the *Journal of Borderland Research* - this time it would stick.

Early in 1995, Tom decided to move to New Zealand (his wife was from there) and he was turning over the keys of the entire publishing operation to me. Thing was, he was selling the property on which the foundation rested. So, I was tasked with finding a suitable place to continue business. It didn't take me long to find that place in downtown Arcata, California.

Arcata is a strange little town located on the Pacific Coast some five miles north of Eureka, California. Arcata, at the time, had more used bookstores in its downtown area than most metropolitan cities had in totality. There was Northtown Books, Arcata Books, the Tin Can Mailman, The Book Exchange, Wiley's Paperback Books, The Rookery, and the this and the that bookstores, many littered with used textbooks from Humboldt State University on detour from the HSU Bookstore on E. Laurel St. From my visits to Arcata, I felt it was the perfect place to relocate Borderland Sciences' publishing company. The stores on the Arcata "Plaza" seemed progressive - there was a futon store, a record store, a health-food store, and home-brew office supply store. Just off the plaza was Los Bagels, The Arcata Co-op, and Wildwood Music. And, the plaza was loaded with several bars for the students of HSU. I secured an office on the backside of the plaza area whose front door opened into the alley across from Marino's Bar. It was a small, two-story office, and the plan was to have the catalog mail-order part of BSRF on the ground floor, and the publishing office upstairs. The office property was owned by a woman - a city council member and later mayor of Arcata. She was an absolutely brilliant woman, who has done much for the town, and she still is. She rented the space to me, I suspect, purely on intuition, and soon thereafter I was moving everything from Borderlands in Southern Humboldt County - the old books, the office machines, and the computers, North, to Arcata.

This part of the story is just history - again - as I remember it. History is important. History is at once a tale of what happened from the perspective of the historian, but may or may not be pure embellishment from the perspective of the reader. Certain readers will question this history based on their own knowledge of the retelling, and from their own research into the matter. The exoskeleton of what is the historical organism, may not jive with the internal organs of its retelling - all historical accounts accurately follow the remembered pattern of those who are its creators. They will be believed or rejected by many who are presented with it, and while this history may be derived from a hostile fanatic wedded to the truth of his own fringe remembrances, it would probably be healthier to accept them as fictions, rather than truths.

Not long after Borderland Sciences found its new home, the kooks started appearing out the very fabric of Arcatatown. I realized after a very short time, that I would have to hire a few people to run the mail-order operation while I worked on the Journal. At first, it was just me - dividing my time between editing the magazine, doing the daily mail, and grappling with the various research projects I'd chosen to create, or recreate based on the old manuscripts I was reading. These old treatises were captured in time in the early years of Borderland Sciences, and written by "Borderlanders" - authors of the kooky science I had now fully embraced. I skipped over the stacks of UFO articles - Borderland Sciences published the first modern encounter with a Flying Saucer in 1946 - a full year before the national media championed the now-famous Kenneth Arnold sighting. These articles written about sightings of UFOs didn't offer anything in the way of experiments that could produce the reproducible results I was looking to fiddle with. I gravitated more towards the early electrical experimenters such as Nathan Stubblefield, Nikola Tesla, and Michael Faraday. I was also keenly interested in all the arts of dowsing - water witching with forked twigs, L-Rods, and a curious invention called the Cameron Aurameter. I chased after the Orgone Accumulator and Cloud Buster of Wilhelm Reich, the weather engineering of Trevor James Constable (it was an article I wrote on the methods Trevor used in his weather modification that inspired Tom to hire me back a second time), and the Pyramid Energies elucidated in the work of Patrick Flanagan. I was also researching the science of Radionics as presented in the works of Ruth Drown, Albert Abrams, and George Starr White. Nowadays, you can easily research these borderland architects online - back then the research was done by reading the print copy of whatever might be available based on the arduous task of archaeological digging in libraries and used bookstores. Fortunately for me, the Borderlands library already held these treasures in abundance.

The first hire was for someone to run the mailroom. We sold books through the mail, so I had to have someone take over the daily orders, package them up, and mail them out to the customers - a rather simple job. I did no advertising for this position - word-of-mouth seemed to spread quickly, and there were all walks of downtown Arcata lifeforms showing up at the doorstep of the new office looking for work. I knew little of Arcata and its subjects save for what I had been told by the few locals I'd met, and I soon found out that the town was a veritable bastion of kookdom. The auditions for the job were absolutely freakish. After 20 or so interviews, I settled on hiring an older gentleman named Gaylord. Gaylord hailed from London, England, and maybe it was the accent that convinced me to hire him over the other applicants, or maybe it was simply that he didn't look like he had just slept in a dipsy-dumpster. Gaylord went to work the following day, and all seemed well - I was able to delegate the mailroom tasks and spend more time doing the things I preferred. I gave Gaylord a key to the office so he could start work early, and for a short time things went splendidly.

About 4 weeks into the job, Gaylord started fucking up. Packages of books sent out to customers were returning to the office frequently. Calls came in from customers who never received their order. On one particularly messy mailroom morning, I confronted Gaylord about his lack of performance, and noticed a smell that was not unlike isopropyl rubbing alcohol coming from Gaylord's breath.

"Have you been drinking? For fucksake, it's 9am."

"Yeah, so what?"

"Well, who the fuck told you it was ok to drink on

the job?

No answer. Gaylord turned away and continued stuffing a padded envelope with books.

"Ok, you're fired."

No response.

"Dude, you are fired, get the fuck out!"

Gaylord acted like I wasn't even there. Fucking British. I went back upstairs and Gaylord finally left at 5pm. I guessed he simply wanted to finish out his last day of work.

The next morning I arrived at the office and Gaylord was already there like nothing had happened. Dammit, I forgot to get the key from him.

"Ok, look man, I fired you yesterday morning. Why are you here?"

No response.

"Alright, if you don't leave now, I'm calling the cops."

Gaylord ignored me and placed a package on the scale. I grabbed the downstairs phone and called the Arcata police department. I was standing only 3 feet away from Gaylord while I explained to the police dispatcher what was going on. Gaylord didn't flinch and kept packaging books. What a nutcase. And I could tell he was half in the bag already - the ripe alcoholic effluvia emanating from his pores was more intense than the day before. The cops showed up some 15 minutes later, and Gaylord was escorted out of the office. He didn't say a word. The cops handed me his key to the office, and told me to call them if he caused any further trouble. I never saw him again. About a month later I received a registered letter from the "Department of Workforce Development" requesting information on Gaylord's termination. Gaylord had the audacity to file for unemployment. I called the Workforce Development office and spoke to Gaylord's case handler.

"So, Mr. Theroux, why was Mr. Gaylord Fuckup terminated?"

"Well, he was drinking on the job early in the morning, and it was affecting his performance."

"Did you have it posted in your office that it was unacceptable to drink on the job?"

"Posted where?"

"Well, on the wall in a place that all employees can easily see."

"Are you kidding me?"

"No, Mr. Theroux, you have to have it clearly posted."

Gaylord was awarded his unemployment claim, and

I experienced another fine example of the chronic stupidity of bureaucracy.

Eventually, I was able to hire another mailroom manager, a bookeeper, a managing editor, and several lackeys to do various jobs around the office that I had no interest in doing. And, there was always a regular cadre of kooks hanging around the office, trying to suck the life out of me while I was working. It was my own damn fault for letting them in.

IN THE OFFICE

Every day around noon, this homeless guy named, "Jaxter" would come by our publishing office to read from the Borderlands library. Jaxter lived out of his backpack, and carried an army of attachments to it, some for sleeping, some for eating, and some for his portable laboratory. Jax called himself a "sensitive scientist" and he insisted he could detect earthquakes long before they would happen. His visits always started with a story of the previous night's events - many were about him getting beat up in the park - by "government agents disguised as hippies." Jax had something new physically wrong with him nearly every day, and appeared "beat up," his face and arms displaying a number of scratches, abrasions, and bruises.

Jax always carried an old plastic gallon milk container that he had refilled with an antibacterial home-brew of colloidal silver. Colloidal Silver is a medicinal broth consisting of silver particles suspended in liquid, and possesses anti-bacterial properties. Jax used it for everything, but mostly for an hideous case of athlete's foot he had, the kind that rotted completely through his tennis shoes. He maintained that it was not an athlete's foot fungus, but a rare case of an obscure, government-engineered flesh-eating bacteria, and that the silver concoction kept it in check. He'd pour it directly into the sides of his shoes while he was still wearing them, and the rotting white leather had been stained an achromatic black color that rose up the sides of the sneakers like an iron mould. I figured it was the best among the many experiments he had going - it was distastefully entertaining.

Jax not only wore a tinfoil hat, but full tinfoil body armor underneath his clothing that he said protected him from the "government satellite mind control beamos." The body armor was of his own construction, which he stated had to be renewed periodically as it became saturated with microwaves. The tinfoil hat, he actually purchased from a mail-order catalog that sold such things. Surprisingly, Jax didn't smell bad, so I let him come into the office and read as long as he didn't bother anyone actually working.

Arriving to the office late one day, I spotted the unmistakable "foot-soldier" contour of Jax walking down the sidewalk. I honked the horn at him and waved as he began crossing the drive into the parking lot, but because of his failing eyesight, he didn't recognize me through the sieve-like lenses of his weather-battered coke-bottle glasses. He sneered at me, and slowed his gait down so that I had to wait for him to cross the entrance. He was in full regalia - backpack cluttered with all the attachments - it was obvious he had just awaken from his slumber in the park, and was on his way to the local CO-OP market for his usual ritual breakfast. This consisted of a quart of soy milk, and a loaf of some pre-packaged manna bread - the kind that's tightly secured in a fancy box. He said the bread had everything you needed to stay healthy - apparently he wasn't eating enough of it.

When Jax finally passed the parking lot drive, I started to pull in slowly. I watched Jax as he marched away, and then I noticed something unusual hanging by a long thread from the side of Jax's backpack. It was swinging back and forth to Jax's step. I stopped for a second to look closer. The thing on the end of the thread began ascending rapidly toward the nape of Jax's neck. It paused just long enough for me to see that it was a giant spider, one of those yellow garden varieties of the Orb Weaver family, and about three inches in diameter. I thought about rolling the window down and yelling to him, but after considering his nasty response toward my need to park the car, I let it go. About an hour later, Jax showed up at the office.

"Hey guys..." Jax had a distinctive knock, and we always knew it was him before he opened the door.

"Hey, Jax, how's it going?" I said.

"Not so good. Take a look at this." Jax pointed to a long white bandage attached to the left side of his neck.

Upon closer inspection, I noticed that the bandage looked a little odd.

"Jax.... ?"

"What?"

"Is that a feminine hygiene product taped to your neck?"

"Well, yeah, they're called sanitary napkins for a reason. I always have a few on hand for injuries."

"Oh, ok. So what happened?"

"A spider bit me."

"No shit?" I said.

"Yeah, it was large and I've never seen one like it before. I think it may have been planted." Jax sprayed something yellowish from a squirt bottle onto the bandage. "I'm keeping the wound impregnated with my mixture of colloidal silver to avoid infection."

"Damn, Jax, that sucks. What makes you think it was planted?"

"Well, there was something weird about it. After it bit me, I slapped it off my neck and it just crawled away. It looked mechanical, not like a real spider."

"Hmmm, interesting. Where would something like that come from?"

"You know. It's the same with all the other stuff. They know their targets and they have to keep us in pain so that we can't function normally. I've heard and seen all of this before. It's nothing new to me. Hey, instead of reading today, I was wondering if maybe you had some work for me to do? I've got a new experiment I'm working on and I need some cash to buy some new batteries."

"Yeah, I might be able to come up with something. What are you working on?"

"It's the earthquake detection thing. I'm using this

old transistor radio to detect the signals - the precursors you know." Jax pulled out an old Panasonic pocket radio from his backpack. "I just tune it in between stations where there's white noise and I can hear certain sounds that come just before an earthquake. I've been using it for a couple of weeks now, but the batteries went dead last night."

"Wow. You ever been able to predict a quake before it happens?"

"Well, I predicted the little one that happened last month, but that was before I had this working. That was when I could still hear the sounds in my head, but after they put up the microwave tower over there on top of the Bank of America, I had to start using this."

"Getting any sounds lately?"

"No, nothing much. I don't think there'll be one for a while based on the quality of the sounds I'm getting."

"Well, that's good. I hate earthquakes. So, let's see, how would you like to put address labels on these envelopes? There are about 150 of them. Shouldn't take you long, and I'll pay you 10 dollars."

"Okay, as long as I don't have to lick anything. They put stuff in the glue on the back of stamps you know."

"No, no licking, just peel the labels off the sheet and stick them in the middle of the envelope like this," I demonstrated.

Jax sat down and went to work on the envelopes and I went back upstairs to my desk to work on the Journal. About two hours later I came downstairs to see how he was doing since I hadn't heard from him.

“I have to go,” said Jax. “I got half of them done so you can just pay me five bucks.”

I gave Jax five dollars, and he suited up and left. I started looking through the envelopes Jax had done. Of the ones he'd actually finished, only three of them were done correctly. Most of the address labels were on upside down or canted at a 45-degree angle. Some had only half of a label attached, the remaining portion still affixed to the sheet they were printed on. Almost all of them were blotched with dirty fingerprints.

That night, an earthquake measuring 6.3 on the Richter scale struck the local area. We didn't see or hear from Jax for some time after that.

THE SECRET LIFE OF PLANTS

Not long after I was able to hire the right people to take over the business duties of the office, I dove straight back into research I was working on - research I would be writing about for inclusion in future Journals. In this case, it was plant consciousness. As I stated earlier, I was quite fascinated by the experimentation in the book, "The Secret Life of Plants," subtitled, *A Fascinating Account of the Physical, Emotional, and Spiritual Relations Between Plants and Man,* by Peter Tompkins and Christopher Bird, and I began conducting experiments based on what little I could glean from the volume. The book, first published in 1973, relied heavily on the research work of one Cleve Backster, an interrogation specialist for the Central Intelligence Agency. In the 1960s, Cleve started hooking plants up to the polygraph machine he used in his CIA interrogations, and this would ultimately lead to his theory of "primary perception" where he claimed that plants "feel pain" and have extrasensory perception (ESP). This discovery of his was widely reported in the media at the time, but was outright rejected by the scientific community. His research seemed a perfect match for the stuff of the *Journal of Borderland Research.*

The most fascinating research detailed in the book was the work of L. George Lawrence, a Silesian-born astrophysicist, who had taken the research of Cleve Backster and improved upon it. So, a polygraph machine uses the fluctuations in galvanic skin response (electrodes are hooked up to the fingers of a subject thought to be lying) and the minute differences in voltage appear as squiggles on an attached strip-chart recorder, not unlike the squiggles one sees on an earthquake seismograph. When you are lying about some fucked-up shit you did, the squiggles will appear on the chart recorder as wild fluctuations, rather than the calm displays of a normal recording. The innovation Lawrence introduced into the equation was transforming the squiggles on the paper into live sound. So instead of interpreting reams of strip-chart paper after the fact, you could now listen in real-time to what was happening with the subject you were interrogating. Lawrence understood the value of his discovery, and immediately adapted this to Backster's plant research - he could now hear what the plants were thinking - as it was happening.

I needed to track down this L. George guy, and find out what he had actually discovered. There were only a couple of articles I was able to dig up from the Borderlands archives - Borderland Sciences had an extensive file system in hard copy on all of the scientists and their research they had ever come into contact with, and there *was* a file on L. George Lawrence.

Unfortunately, the file only consisted of one article - but it was significant - the article was a reprint from *Popular Electronics* entitled, "Experimental Electro-culture" from February, 1971. The key to finding Lawrence was a reference in the article to the "Ecola Institute." We had also published an article by L. George Lawrence in the *Journal of Borderland Research*, Vol. 29, No. 4, July-Aug 1973. This article also made reference to the Ecola Institute. But, in this article entitled, "Interstellar Communications Signals," Lawrence made reference to receiving communications from outer space - from biological materials. Lawrence claimed he accidentally received intelligent signals from the region of the constellation Ursa Major (The Big Dipper) while experimenting with his plant detectors!

Using my sleuthing skills, none of which were acquired while working with the private investigator, I found a website on the internet named “Ecola Services,” a pest-exterminator company that used a curious electromagnetic device based on the work of Nikola Tesla. Somehow, I thought this had to be connected to L. George Lawrence. I called the phone number listed on the site and started asking questions. I found out I was actually talking to L. George Lawrence’s niece, Susan, and that the company was based on his invention of the electromagnetic pest eradicator. Poor L. George died some years ago, but his works lived on in this company, and it was called the “Electro-Gun Termite Treatment.” From Plant Consciousness to Extraterrestrial Communication to Termite Extermination? This was getting wackier by the minute.

Susan stated that she had all of L. George’s old notes on his research, and some test equipment that she’d be happy to give me if I were to pay her a visit. The company was based out of Las Vegas so it would be quite a road trip, but well worth it. Three days later, I was in Vegas talking to Susan in person. After seeing my enthusiasm for L. George’s work, Susan gave me everything - test equipment, research papers, and two things she hadn’t mentioned on the phone. One was an 80-page patent of Lawrence’s entitled, "Methods and Receiver for Biological Data Transport". The second was a 300+ page unpublished book he had written - “Galactic Life Unveiled,” all about his discovery of interstellar communication. It was like winning the lottery.

I loaded up my Isuzu Trooper with oscilloscopes, military radios, boxes of books and research papers, and all kinds of other indecipherable test equipment. The truck was loaded from top to bottom. The truck was so laden with what was left of L. George, it would only achieve a top speed of 25mph driving up California 190 through Death Valley - a very long 7% grade. While chugging along at this snail's pace, I decided it might be interesting to camp overnight at the famous "Racetrack Playa." The Racetrack Playa in Death Valley National Park is a dry lake bed that is most widely known for the stones that leave long tracks as they slide across the lake bed floor. No one has ever seen these rocks move (some can weigh up to several hundred pounds) so it is still quite a mystery. The wind on the playa is like nowhere else on the planet, and setting up camp really sucked. As I was trying to set up my tent, a wicked gust blew up and carried it away like a rolling tumbleweed - it was gone in seconds. I would be forced to bivouac in what little room was left in the interior of the truck - the driver's seat. Initially, I had the idea that I would set up video cameras on tripods - so that they might capture the mystery of these sliding stones. I set up two cameras each entrained on rocks that had showed visible signs of having previously moved. I fired up both of the cameras - set to record for 6 hours, and retired to the cramped Isuzu shelter. Sleep was barely possible, but I did doze in and out. On the out times, I would look at the stones to see if I could detect movement. Nothing was happening. Finally, the sun rose over the mountains, and it was time to check things out. The cameras had been knocked down, and both were now broken. I looked at the damn stones, and they hadn't moved an inch. What a complete waste of time - the mystery of the Racetrack Playa's sliding stones - was still a mystery. There have been some recent explanations that have become accepted as fact about the phenomenon...but, to this day, no one has witnessed or filmed the stones in action.

When I finally arrived back in Arcata, I directly dove into all the research of L. George Lawrence. I originally set out to recreate his work with plant consciousness, but now I had to confront his ideas on interstellar biological communication.

From the notes I had acquired, Lawrence writes:

"The possibility of life beyond Earth tends to evoke partisan emotions denoting two extremes (1) that life is unique to Earth, and (2) that it is ubiquitous and abundant throughout the universe. Hints about life elsewhere have been vague, including "organised elements" in meteorites, Martian infra-red spectra implying organic molecules, and the rare non-random radio signals whose reception was claimed by Tesla and Marconi. Yet a psychological readiness to accept extraterrestrial life has emerged to a most profound degree.

Divorced from its science-fiction background by Project Osma in the early 1960s, the topic is now formally called "communication with extraterrestrial intelligences" (acronym CETI). It was the subject of a top-level international conference held in September 1971 at the Byurakan Astrophysical Observatory in Soviet Armenia. The meeting was sponsored by the United States and Soviet Academies of Science. Similar interests prevail in private study groups both here and abroad.

Such programs must be welcomed in behalf of all mankind. For interstellar communication with other, highly advanced cultures might well determine human progress and survival. Thus, by inference, CETI emerges not only as the greatest adventure of mankind, but also poses a colossal challenge to systems engineering.

It remains an unfortunate fact that the majority of ambitious CETI programs continue to be aligned with the somewhat obsolete theories and hardware of radio astronomy. This is a serious drawback for several reasons: (1) local inadequacies in interstellar communications technology and humanness are imposed upon a given extraterrestrial society (prompting it to lose interest in us), and (2) we must be aware of the pre-eminent possibility that Type II or III civilizations -- whose worlds were old when Earth was young -- no longer employ electromagnetics-based communication technology due to obsolescence. The latter assumption is defensibly supported by the fact that terrestrial radio astronomy, in its 40 year long history as an electronics-based science, has failed to detect intelligent, non-Gaussian signals from deep space."

Lawrence was certainly on to something big here, but would it really hold up to the scrutiny of the reproducible results I sought? The main problem I have with L. George's work regarding extraterrestrial communication, is that while he stated that he accidentally discovered intelligent signals emanating from the direction of the constellation Ursa Major, how could he know that these signals originated directly from The Great Bear to earth, or could they be coming from somewhere in the space between? Ursa major is roughly 18.6 megaparsecs (60 million light years) away from the earth - there's a lot of space in between the two. According to conventional science, it would take the light from those stars 60 million years to reach the earth, and if the speed of light is the accepted limit with which we can communicate, then who cares about possible messages that took 60 million years to get here?

I accept L. George's theory, that the electromagnetic spectrum is far too limited with regard to interstellar communication - one would certainly have to transcend that if ET communication were to be at all viable. What I didn't accept was his methodology in acquiring his claimed communication. How does one direct, or rather point such a biocomm device in any direction? How would one know whether or not the biocomm medium was receiving or transmitting? We're just not there yet, and it will be some time before the science of earth catches up with a more advanced science that can facilitate instantaneous communication over vast distances. Oh, I forgot to mention that L. George was a good friend of George Van Tassel - Van Tassel allowed L. George to set up his lab in the vacated apartment underneath Giant Rock!

The reading of L. George Lawrence's research papers, unpublished manuscripts, and patents leaves one as dumbfounded as to what he was doing, as the reading of Ed Leedskalnin's discoveries on the secrets of the construction of the pyramids. Too cryptic, and I suspect, that is exactly how kooks have intended their works to be interpreted. Critics will say that I just don't possess the ability to understand these works, and that there will be someone out there that can. Good luck motherfuckers.

While I am still wholly enthusiastic about this endeavor, and continue to develop new technology to facilitate this work, there are too many questions yet unanswered. A higher consciousness is needed for this kind of dedication - dedication to research such as this. The role of the kook may just be the role of the visionary, or seer - that hallucinatory component of madness which is married to ingenious discovery. I understand there are medications that can control this behavior, or rather do they simply suppress it? To true visionaries and seers, medication is a poisonous mixture of science gone wrong, and a suppressible agent designed to kill off the creative spirit. To step up to the next level of this research is to narcissistically assume you have the answers to its solution. Or maybe you do have the answers. It's all in the interpretation. Genius is never won overnight, it is usually honored long after the subject of its scrutiny is dead.

I had converted a good portion of the office of Borderlands into a plant consciousness research laboratory. The device that was hooked up to the plants ran 24/7, and while it was initially considered "irritating" to the staff, they eventually found it somewhat interesting, and quite amusing. Following L. George Lawrence, I'd fabricated an electronic device that would translate the galvanic responses of plants into an audible wall of cacophonous sounds that permeated the entire workspace.

Enter Princess Leia. Leia was a local real-estate agent I'd met one morning at Los Bagels, my early *ritual de lo habitual* coffee shop on the North end of Arcata.

We struck up conversation, and small-chatted about what was going on in our lives, and what wasn't. I invited Leia to have a tour of my office to check out the crazy shit we were doing, and the next day, the princess appeared at the door. Leia was 6' 2", blonde, and what I otherwise considered a rather a plainfest specimen of the fairer sex. But, Leia had a bit of crazy going on. My kook magnet was still in full-on attraction.

Normally, the daily plant sounds of the office were rather unexciting - they simply remained steady in their Gregorian song throughout the day. This was to be expected since we weren't torturing them as did other researchers during the course of their experiments. One of the most famous examples of plant torture Cleve Backster conjured up, was to dump a ladle of water full of sea monkeys into a pot of boiling stew, all the while monitoring the plant's response to the event. He murdered a bunch of brine shrimp to see if the plants would notice. They fucking noticed alright, the strip chart recorder hooked up to them responded with flailing seizures of pen on paper.

This is exactly what happened the moment Princess Leia stepped into the Borderlands office. The sounds coming from the apparatus hooked to the plants jumped immediately from their usual baseline droning, to a wild screaming - with no apparent provocation. Leia had no idea what was going on, and asked what the "noise" was all about. I explained to her that the plants were reacting to a "stranger" in the room and that it was not unusual (it was highly unusual since no one else's entrances elicited such an intense response).

"So, explain to me how all of this works."

I had no idea where to begin. In my opinion, having experienced the various reactions of the plants to many people over several months of observations, they were simply in fear of her presence.

“They seem happy and are excited you are here,” I lied.

“Wow! Well, what would happen if you hooked those electrodes up to me?”

“Uh, I dunno, you mean hook this whole electronic thing up to you?”

“Yes, maybe you could write about me in your little newsletter instead of these stupid plants.”

Great, she wanted to be featured in my “little newsletter.” I had to bite.

I carefully disconnected the plant from the device, and applied the electrodes to her index and middle fingers. I was familiar with using such a device as a biofeedback monitor on people - and that one could control the output of the sound by doing certain relaxation exercises. Leia was not a normal subject. After turning the device on, no sound came from it, not even a squeak. I adjusted the sensitivity from the lowest setting to the highest - with no result at all. After a few minutes, a very high pitched tone emerged and slowly descended in pitch to what could be considered a baseline tone.

"Is that normal?"

"Sure," I said, "Perfectly normal. Just try to relax."

"Well, I was supposed to meet with a client about 20 minutes ago. I guess I'll get there when I get there. I feel all tingly."

My kook magnetron was fully engaged.

The basic test to get a reaction from the device after having obtained a baseline tone, is to lightly pinch the subject connected to it. So I pinched her. No response whatsoever. I pinched harder....nothing. Leia seemed to be getting excited, yet there was no response from the device that was hooked up to her. I pinched and twisted her all over her entire body, and still, no response from the tone emitted from the device. It was a completely unwavering tone.

"Michael, this is really amazing, but I have to go now. Call me?"

"Sure, I'll call you."

For some reason, I did call her, that night. According to the biofeedback experiment she insisted I try on her, she appeared to be galvanically, and organically dead. I guess I was intrigued, and had to see if she was really alive. That or I just wanted to get laid. That night we had sex - it was horrible, and I confirmed that she was indeed, dead.

MARGARET THORSEN AND HER FREE ENERGY DEVICE

Free Energy was a topic that was frequently explored at Borderland Sciences. There were always many assertions made by self-proclaimed research scientists of their discovery of new free energy devices, but in nearly all cases they were simply the spurious declarations of nutcases and kooks. In nearly every claim, there was an attendant proclamation of some subversive form of suppression going on, usually by secret government agencies that prevented these scientists from receiving the recognition they deserved. While I have no doubt that the suppression of certain technologies has occurred in the past, most of the cases we came across were the consequence of a very high degree of conspiratorial imagination. So what is "free energy?"

Free energy can be unambiguously defined as a source of usable energy that costs nothing to produce. In the case of many free energy researchers, free energy is more of a simple equation of efficiency. It is a comparison of the work output from an operation to the work input to that same operation. This would be defined as;

Efficiency (η) = Power out / Power in

% η = (P out / P in) x 100.

While this is a simplification of the efficiency

equation, its misinterpretation exemplifies the irrational thought processes of many free energy researchers in their quest to be recognized. The story below is a classic example of this:

———————

Early one morning, there was a knock at the door to the office. I opened the front door, and barging in, was a very large woman, so large, in fact, she had to turn sideways to get through the door.

"My name is Margaret Thorsen and I'm here to show you free energy."

"Did you have an appointment?" I asked.

"Um, no, should I come back some other day?"

"No, c'mon in. So, tell me more."

"Well, my father invented this free energy device which I got in my trunk you know I drove all the way from Galveston and I didn't sleep much because I knew that I was bein' called to come here and show Daddy's invention and I can come back another day if you think that would be better."

She appeared to be her late 50s, and weighed in the neighborhood of 350 pounds. Due to her intense, fidgety mannerisms, I suspected she was high on something.

"Well, why don't you go get it, and we'll have a look." A couple of my co-workers began chuckling under their breath.

"Well, see, it's kinda heavy. I'll need some help carryin' it and I'm parked in the grocery store parking lot about 2 blocks away so it will be too heavy to carry that far."

"No problem. I'll help you."

We walked to the parking lot and the woman stopped at a brown, rusty 1973 Oldsmobile 88 with Texas plates. It was covered in filth and had finger and hand prints all over the exterior.

"Sometimes this here trunk key don't work too good." She fidgeted for about 5 minutes trying to get the trunk open.

"Can I help with that?"

"Naw, I can get it. You gotta have a certain special touch with it."

"Yeah, I can see that."

After a couple more painful minutes, the trunk popped open and there it was....the "free energy" device. I lifted it from the trunk and carried it the two blocks back to the office myself. It was composed of 2 coils; made up of sloppy 12 gauge windings wound transverse to another length of 28 gauge windings surrounding a large iron block, and mounted on a slab of plywood. It looked like a transformer.

"So, your father invented this?"

"He did. He showed it around to a lot of engineers who couldn't never figure it out and they saw it working just like this one but this one I made from my Daddy's instructions he left me when he died last year."

I set the "device" on the table in the office. It had a cheap RadioShack VOM nailed to it with the probes clipped on to one end of the secondary coil.

"Ok, fire it up!"

"You got an outlet close by so I can plug it in?"

"You have to plug it in? I thought it was free energy?"

"Oh, it is free energy alright, but you gotta get it goin' first and it has to be plugged in for it to get going and then it goes higher than what you think it would on the meter."

"Ok, but you realize this is just a transformer, right?"

"Oh, no, this is all Daddy's invention he made when he was a kid and then he made more until he got it right. My Daddy was a inventor and a genius."

I had one of my associates plug the thing in just in case something went wrong. The meter jumped to 500 volts.

"Isn't it amazing? Now I'm gonna explain what is happening with the science of it with the meter that shows it is 500 volts and that's all free minus the 120 volts that comes from the wall there and if you subtract that you get around 400 volts which is all free energy."

"Um...can I explain this to you? This is nothing more than a step up transformer. It isn't free energy."

"What are you saying?"

"It isn't free energy. It's an old invention - Michael Faraday in the 19th century? Ever heard of Nikola Tesla?"

"Oh yes, Telsa [sic] had free energy too - the Telsa Coils he made lightning with 'em and powered all of New York city with them until the government took it all away from him and burned his house down. I think they killed him too."

"Well, that's not quite how the story goes. Anyway, I want to show you something." I walked over to a bookshelf in the corner of the office and retrieved a neon sign transformer.

"This is a step-up transformer. But if you look at the face plate on it, it says 120 volts input, and 10,000 volts output at 23 milliamps."

"That ain't possible. Where did you get that?"

"It's a common neon sign transformer. This one is from 1948."

"What are you trying to say? That my Daddy's invention ain't no good? I take this all over and there's been a lot of engineers that looked at it and they can't figure it out so you don't go tellin' me that it ain't free energy. Are you working for the government too 'cause if you are I'll tell a lot of people what you're doing and that you don't really want no free energy."

"That's fine ma'am. It's time for you to go now."

She squeezed out the front door with "Daddy's invention" and I never saw her again. A week later I walked past the supermarket parking lot and her car was still in the same spot. I could only hope that she and her "free energy" device somehow made it home.

ERIC DOLLARD, BOLINAS, AND BACK TO THE HIGH DESERT

If anyone could recreate Nikola Tesla's research, that anyone would be Eric. Not long after I started working with Borderlands, I discovered and helped publish some of the new works of Eric Dollard (Tom Brown had discovered this Tesla genius). There were numerous articles, books, and videotapes by Eric - all produced by Borderland Sciences, and all of them were about his discoveries on the work of Nikola Tesla. Tesla was the subject of many a Borderland publication, but Eric's work on Tesla seemed so much more academic, learned - and accurate.

Eric's appearance, save for the t-shirts, jeans, and no shoes he wore, immediately hearkened one back to 1900 - the long, full-faced beard that emerged from the bottom of his eyelids and continued all the way to the base of his neck; the round wire-rimmed glasses, and the bald top of his head conjured the image of a turn-of-the-twentieth-century scientist. Every word and sentence he spoke was crafted with articulation and elocuted with precision. No "and-uhs" or prefaced sentences with "basically," "actually," or "technically speaking," were ever uttered in his delivery. I was fascinated by his apparent genius, and needed to know more.

My first face-to-face encounter with Eric was in the town of Bolinas, California. Bolinas is located about an hour north of San Francisco, just up the coast near Stinson Beach. Nested on the southernmost tip of the Point Reyes Peninsula, Bolinas is on the western side of the San Andreas Fault, the side that's pulling away from the continent. Kevin Opstedal in his article, *A Literary History of the San Andreas Fault: Bolinas Section* wrote,

"Perhaps this geological fact has had an effect, on some cosmic level, upon the consciousness of the Bolinas citizenry. Since the early seventies the media has portrayed Bolinas as a home for the odd, the wacky, and the certifiably strange. While it's true that the place attracted more than its share of bohemians, hippies, visionary sandal-makers, and mystic organic farmers, it is unfair to view the place, as one Marin County supervisor did, "an asylum run by the inmates".

Bolinas was also home to the Marconi/RCA Wireless Station, and this is why Eric had made the strange town his home. The Marconi wireless station was the foundation for the most successful and powerful ship-to-shore communications in the early years of radio, and Eric was keenly interested in this. Eric stated that the site's technology wasn't that of Guglielmo Marconi, the 1909 winner of the Noble Prize for Physics, and acclaimed inventor of radio - but the invention of the true inventor of radio - Nikola Tesla.

When I first arrived in Bolinas, Eric escorted me to the old Marconi station and gave me the grand tour. It was a well-narrated tour, and Eric seemed to know more about the place than anything I had ever read about it. After the tour, Eric lead me back to his homestead. This was nothing more than a lot at the end of what appeared to be a residential neighborhood - there was no structure on the lot - it was simply a garden of military radio equipment. Eric cracked a couple of shitty domestic beers and handed me one. For another hour, I listened to Eric go on about how he was shafted by this and that group, and that all were out to get him. He drank four beers to my one, and his tone got nastier about the local politics after each twist-off. It was nearly 11pm, and the radio receivers were blaring a cacophony of shortwave and ham radio conversations for the entire neighborhood to hear. I said I had to slab for the night, and asked where I was supposed to sleep.

"I sleep in my car and listen to the radios, so I guess you're on your own."

Eric had removed the passenger seat from the old Toyota Corolla he bedded down in every night - in its place was large WWII military shortwave radio. Eric kicked the driver's seat back, and slept - or slept and listened (later I would adopt this habit on cross-country road-trips - minus the removal of the passenger seat - it felt like a great deliverance from the uniformity and commonality of normal camping, and the radio's constant banter was a better companion than most humans).

That night, I retired to my Isuzu Trooper for a very restless sleep. In the morning, Eric brewed up a mudsludge of cheap cowboy coffee over an open fire, and I continued listening to how he had been fucked over by everyone he'd ever met.

Years later, after Eric had yet again been evicted, this time from his lot in Bolinas, I was invited to his other home away from home - the High Desert of Southern California. His forced removal from the

Bolinas lot was rather like a military operation, with the entire staff of the Marin County Sheriff's Department in attendance - including half a dozen bulldozers. I filmed the entire destruction of the property and Eric's ejectment that morning. Mysteriously, the film no longer exists.

Eric was now staying at a strange piece of gravel in the Landers area of Southern California with a man named Walt, not far from the old Integretron site. As I was invited to the place, I drove right up the long driveway to the foot of the main building on the property - a single-wide trailer. I waited in the truck for signs of life for nearly ten minutes - the place appeared to be abandoned. Suddenly, the main door to the trailer flew open, and a white-haired old man in overalls tumbled out, wielding an ancient revolver screaming that I was trespassing on private property and that if I didn't want to be shot, I'd better leave. Great. I was going to be killed by some desert kook, and all I did was accept an invitation to see Eric. Not a moment too soon, Eric himself appeared from around the backside of the trailer and yelled at Walt,

"Don't shoot! He's a friend, don't shoot!"

"Well who the fuck is he?" Walt shouted.

"I forgot to tell you he was coming. He's from Borderlands."

Walt holstered his pistol and revealed a toothless grin.

“Shit, hehe, I almost shot yer dumbass. I thought you was commin’ to take my dogs. You want a beer?”

I got out of my truck and walked toward them. Walt backed up and put his hand on his pistol.

“You better be cool or I’ll fuck you up, motherfucker!”

I seriously did not want to be here.

“Hahaaa! Gotcha din’t I motherfucker?”

“Walt, settle down and go get him a beer,” Eric calmly commanded.

Walt hobbled back inside the trailer and a pack of yapping dogs welcomed his return. Eric brought me into the trailer to sit down at the front table for a beer. Immediately, I noticed the thick stench of dog piss and looked down to discover I was standing in it. Walt returned from the back end of the trailer with three beers, and four miniature poodles in tow.

“Don’t mind the piss, they’re still in training...haha,” Walt said.

I had to get away from this disgusting human, so I said I wanted to see what Eric was working on before it got dark. Eric led me out of the trailer down a sandy path to a small shack several yards away. As I stepped out of the trailer, my footsteps sounded much like the separation of strips of Velcro - created by the sickening sticky sound of Vibram being lifted from its contact to a half-dried pool of dog urine on the trailer floor. Inside the small shack were the usual racks of military communications equipment I was accustomed to seeing wherever Eric made his home.

“This is my new earthquake detection station. That biosensor you sent me is connected to all of this equipment. I think you are really onto something with those biosensors you’ve made.”

These were the simple biosensor devices I created from the works of L. George Lawrence. L. George had invented this idea 30 years ago at Giant Rock, not even 20 miles from Eric’s present location in Landers, and now these biosensors had returned to the High Desert from whence they came. Irony was very evident at this moment in time.

The biosensor device was part capacitor, part organic material and it was designed to replicate the biological activity in the leaf of a plant. The thing was, these artificial biosensors lasted only a few weeks, far less than the leaves of plants, and plants cost nothing and required no construction. I eventually returned to using plants instead of the costly biosensors. Now, years ago, in 1992, when I was living in the Northern Humboldt County California wilderness with no power - miles from any connection to the electrical power grid - I was able to set up experiments recording Very Low Frequency (VLF) transmissions from the earth. VLF is defined as any radio frequencies (RF) in the range of 3 kHz to 30 kHz, and wavelengths from 10 to 100 kilometres. Being so far away from any 60 cycle interference, the signals came in with very little hum from the grid. The apparatus I used consisted of two iron rods separated and driven into the earth and then connected by shielded wires to the audio inputs of a simple Panasonic boombox radio. What I was doing was listening to the otherworldly sounds of the various pings, pops, squeals, and "whistlers" that were coming from the earth itself. I really didn't know what I was listening for at the time, but it was cheap entertainment. I had been experimenting with a number of grounded antennas for radio reception based on numerous patents we had on file at Borderlands from the 1920s and 1930s - patents that clearly stated that radio reception was far better using underground antennas than those situated above ground. And it was true. Radio stations from thousands of miles away came in much clearer than with the use of a conventional above-ground antenna. While the VLF receiver (boombox) was picking up these signals, I was also using the cassette recorder on the device to record the sounds in real time.

What I didn't know was that I would be experiencing three massive earthquakes in the next 24 hours, and that they would all be recorded on my experimental setup. This experimental setup, while still crude, was far more scientific than the likes of what Jax would later describe as his "transistor radio earthquake monitor." Hours before each quake occurred, the intensity of the VLF sounds increased dramatically, and during the actual earthquakes, revealed sounds never heard before on the recordings.

I knew Eric had heard these tapes, and that he had read the article I'd written about my experiments in the

Borderlands Journal. The lab he had now set up in the desert seemed to have emerged from the direct result of the experiments I had initially conducted, even though his setup seemed far more sophisticated. I guess I would have been slightly honored had he even mentioned once that he had developed the research based on what I had previously done, but it was as though he had come up with it all by himself. Typical to the world of the fringe sciences was the rampant borrowing of ideas with no reference to the originator of such ideas, and much to the detriment of this academically unaccepted field. While the scientific misconduct of not crediting other's previous work does appear in the conventional scientific community, it is rare compared to that which occurs in the alternative science collective. But, I would eventually realize it was pure jealousy on my part that Eric could put together a far more advanced VLF research lab. His work on the subject was superior, and I simply focused on trivial bullshit to make myself feel better about it.

Darkness fell over the high desert compound, and I decided it was time to make camp. I pitched my tent about 100 yards from the trailer and resolved to get up early and leave the desert crazies to their brown, dusty, and dismal existence. The temperature dropped to around 20 degrees and sleep was again restless. I tore down camp at 6am, packed my gear in the truck, and excitedly engaged the starter. It turned and turned and nothing happened. The engine would not go. Eric emerged from somewhere and said, "Pop the hood."

Eric started removing shit from inside the engine compartment - first the air filter housing, then the injector assembly, until the very pistons of the damn thing were naked to the swirling grit of the desert morning. My only thought at this point was that I was going to be stuck in this hellhole forever waiting for motor parts from civilization, or that the truck may never run again. Eric worked on the engine for a couple of hours, slapped everything back together, and said, "Try it now."

The fucking thing roared to life like it was new off the dealer's lot. Minutes later, as I was driving north, away from that nutty desert coop, the epiphany of Eric's genius hit me. He may have been a wacky kook of the first order, but he was a brilliant man that could build and fix anything - and he was my personal saviour on that day, delivering me from the evils of the kook-encrusted Mohave Desert.

Walt died shortly thereafter, and as I understand it, Walt had left his desert compound to Eric. I heard that for some reason or another, Eric lost that place too - the sordid repeating story of Eric's life. Just last year, I heard Eric was giving a talk at some energy conference about his work with earthquake detection. His "agent" contacted me to see if I would call Eric - he wanted to use some of my earlier VLF earthquake recordings for his presentation. As all of those recordings were freely available on my website, I told the agent that Eric could just download and use whatever he wanted. I politely bowed out when I was invited to attend the conference...it was not a situation I really wanted to rekindle. And, again, I never heard from Eric or his agent again. I'm sure you are seeing the pattern here.

UNCLE WALTY AND THE HOLLOW EARTH INSIDERS

Borderland Sciences stored a treasure of old works dating back to its inception in 1945. These treasures consisted mostly of printed material, and in the tape-recorded lectures of the past directors. It was a challenge to know just what to do with them. But the most voluminous of the collection consisted of the hundreds of recorded cassette tapes of all of the lectures that Riley Crabb had done while he was director from 1959 to 1985. These ranged in topic from of the Huna Religion of Hawaii, to his Cabbalistic interpretations of the flying saucer phenomenon. The tapes were stored in cardboard boxes in a storage facility, and were rapidly disintegrating. I thought it would be a good idea to offload these tapes to someone who was interested in them, and I placed an ad in the Journal - if anyone wanted them, we would ship them free of charge.

Shortly after the magazine hit the newsstand, I got a call from a man named Michael Elsey. Mike was a recording engineer for MCA Records, and he stated he would be more than happy to acquire all of these tapes - and that he would restore them to their former glory.

In the early 1990s, I began doing recordings of experimental music I'd composed, and started offering these on cassette tape through the Borderland Sciences catalog. The first album was entitled, *Mundus Subterraneus*. It was created with pianos, harmoniums, shortwave radio sounds, and earthquake sounds generated from the very low frequency recording apparatus I had been working with. The title of the album, *Mundus Subterraneus,* was lifted from a rare manuscript by Athanasius Kircher (1665) that describes and details energetic currents from inside the earth itself. This album of music was an early exploration into the dissociation of rhythm, melody, and harmony that would later be realized in my book, *Rhythmic Formative Forces of Music.*

I sent the boxes of Riley's tapes to Mike, and I included a copy of *Mundus* for him to listen to. Mike called me shortly after he received the boxes to tell me he was working on Riley's lectures, but he was very intrigued by my recording of *Mundus Subterraneus,* and as I was to find out later, he was completely infatuated with the idea of its theme.

Mike talked of his research into earth energies and of the concept of the Hollow Earth, and he was rather passionate about the topic. He talked about Dr. Raymond Bernard who was born "Walter Seigmeister" in 1901. Bernard became deeply interested in the works of the Theosophists and Rudolf Steiner's Anthroposophists, especially those works relating to Atlantis and Lemuria. This Dr. Raymond Bernard wrote a book entitled, *The Hollow Earth* which exemplified the idea that UFOs were coming from inside the earth. Bernard also added that the Ring Nebula proved the existence of hollow worlds, as well as his speculation on the fate of Atlantis, and the origin of flying saucers. Bernard stated that the inhabitants of Atlantis took refuge in the Earth's interior before the now-lost city was destroyed in a great calamity. It was the Atlanteans who piloted the flying machines known in ancient India as *vimanas*, and now known in the modern world as flying saucers. After the U.S. bombings of Hiroshima and Nagasaki, Bernard claimed the Atlanteans became concerned that radioactive air might flow into the world's interior, and so many of them emerged in their flying saucers in an act of self-defense. Later, an article by Martin Gardner revealed that Walter Siegmeister had used the pseudonym "Dr. Raymond Bernard", but not until the publishing of Walter Kafton-Minkel's *Subterranean Worlds: 100,000 years of dragons, dwarves, the dead, lost races & UFOs from inside the Earth*, in 1989, did the mystery of Bernard/Siegmeister become well known. Sir Walter Scott is said to have been keenly interested in the tales of hollow earths as well. And, there is another tale from a "Walt" - Walter Map's *De nugis curialium* (Trifles of courtiers). Of Welsh ancestry, Map was a medieval chronicler of the story of King Herla which includes numerous references to subterranean worlds and its inhabitants. So many "Walters" in this tale of hollow earths - it was getting rather creepy.

Mike then told me about his favorite Hollow Earth researcher. Bruce Allen Walton, known affectionately as "Uncle Walty," wrote *A Guide to the Inner Earth* among other books on the hollow earth, that included references to specific locations where one might enter these supposed underground portals. One of those locations was Mt. Shasta, in Northern California. Mount Shasta (Karuk: *Úytaahkoo* or "White Mountain") is a volcano located at the southern end of the Cascade Range in Siskiyou County, California. At 14,179 feet, it is the second highest peak in the Cascades and the fifth highest in California. Mount Shasta has an estimated volume of 85 cubic miles (350 km^3) which makes it the most voluminous stratovolcano in the Cascade Volcanic Arc.

There is a large body of supernatural lore associated with Mt. Shasta. Native stories abound with tales of mythical beasts that lived inside the mountain, and much later, anecdotes recount flying saucers that are hidden within the lenticular cloud-forms that frequent the summit of the volcano. Hundreds of Sasquatch (Bigfoot) sightings have also been recorded in the area. According to some, the underground city of "Telos" exists directly beneath the volcano, and is inhabited by the the people from the Lost Continent of Lemuria, who have been living underground for the past 12,000 years, and due to their isolation from the surface population, have created a civilization of peace and abundance, with no sickness, aging, or death. This underground city boasts an inner central sun, complete with oceans and mountains much like the surface of the earth. Mt. Shasta was a truly a place of mystery and imagination, and Mike Elsey said we needed to investigate.

Shasta was only about three-and-half hours inland from coastal Arcata, so it would be an easy road trip for me. For Mike, it would be almost nine hours of driving. Mike said that Uncle Walty stated in his other book, *Mt. Shasta, Home of the Ancients,* that there was a portal to the Hollow Earth near McCloud, California, on the eastern side of the mountain - more specifically, an area called “Elk Flat.” Walty said that the caves of the Elk Flat region “lead to an underground village where a race of tall beings live and move.” Telos! Mike had obtained some forest service maps of the area - topographical maps - that indicated two caves just west of Elk Flat, and that would be our destination. We drove for several miles on the dirt road approaching the area, and finally stopped to set out on foot. GPS at this time was expensive, and accuracy was left to the military alone, so our cheap camping compasses would have to do. An hour or so into our hike through the forest, I spotted a strange rise in the terrain. According to my crude calculations, the main cave should be just on the other side of the rise. At the top of this rise, there was an indentation mostly covered with brush - it was the entrance to one of the Elk Flat caves. The entrance to this cave wasn't the typical hole-in-the-side-of-a-mountain type of cave, it was just a fucking hole in the ground. The entrance was so concealed and remote, I thought no one could have been there for decades. Mike and I descended into the cave entrance, and at about 20 yards into it, we were immediately struck by its vastness. The ceiling of the cave had to be at least 20 meters high, and its width easily equaled that. It was cathedral-like in its structure, almost as if its illuminated spires were man-made. Oh yes, we had light sources. These were cheap incandescent headlamps procured from some REI-type outlet, and running on double-A batteries. They were supposed to last for four hours. According to Uncle Walty, it would only take two hours to reach the lighted village within the interior of the mountain, so we were confidently prepared for the adventure.

An hour or so into our excursion into the interior of the hollow earth, we came upon an ancient artifact of spiritual significance. Deftly situated upon a cairn of small rocks, was a "God's Eye." A God's eye is a yarn weaving (Ojo de Dios - *Eye of God* in Spanish) and is woven with yarn and wood, often with several colors. The weaving of an Ojo de Dios is an ancient contemplative and spiritual practice for many indigenous peoples in the Americas. Upon closer inspection, this particular "Ojo de Dios" was made from popcicle sticks and several colors of acrylic yarn. No matter, maybe some indigians left it there for us to find. Another hour deeper into the cave, we realized it was significantly narrowing. We soon thereafter reached an impassable end to the cave, but it was here that we were greeted by the most alien of artifacts. Hoisted at the ends of several vertical sticks were the soiled remains of women's panties, monuments to all those who would reach the inner depths of this mystical entrance to the hollow earth. Mike sighed deeply, and we immediately began our return from whence we came. It was a particularly uneventful and silent resurfacing.

Most mythological stories seem to fail miserably in the face of vigorous confrontation. The stories are still great, and at least, they push the envelope of the usual mediocrity of our modern imaginations toward exciting investigation. The outcome is less important than the realizations had upon the journey. This is the main *raison d'être* this book was written - to recount the voyage, not the end result. And what a long, strange trip it's been. Most of the kooks recounted in these pages were all good people. They mean no harm, being kooks - they simply have wild imaginations that sometimes get the best of them. But, there is another order of kook that embodies all that is profoundly immoral and malevolent in this world, and in the next chapter, we'll investigate one of these evil bastards.

WILHELM REICH, TREVOR CONSTABLE, AND JOEL CARLINSKY'S BONFIRE OF INSANITY

Soon after my first firing from Borderland Sciences, I was back in North Dakota, working at a local music store, and playing my ass off. I was the guitar player in a three-piece blues band and having a good time. By day, I moved pianos and I eventually began tuning them for income. But, this amusement was by no means fulfilling, and the vacuum that had displaced my former life grew to a point where I felt it necessary to fill the space with experimentation and writing. My first article for the Borderlands Journal was *Weather Engineering on the High Plains* - a discourse on several experiments I had been conducting in relation to the work of Wilhelm Reich and Trevor James Constable.

Wilhelm Reich, (1897 - 1957) was an Austrian psychoanalyst, a student of Sigmund Freud, and one of the most radical figures in the history of psychiatry. He was the author of several influential books, most notable to me were, *The Discovery of Orgone, Volume 1: The Function of the Orgasm*, 1942, and *Cosmic Superimposition: Man's Orgonotic Roots in Nature*, 1951.

Reich stated he had discovered the life force - the primordial cosmic energy - and he called it "Orgone energy" or "Orgone radiation," and the study of it "Orgonomy." He argued that it is in the soil, in water, and in the air, its color is blue or blue-grey, and that humankind had split its knowledge of it in two: "ether" for the physical aspect and "God" for the spiritual. The color of the sky, the northern lights, St Elmo's Fire, and the blue of sexually excited frogs are manifestations of Orgone, he wrote.

Sometime around 1940 he began to build an enclosed device that consisted of alternating layers organic (wool) and inorganic (metal) materials much like that of a capacitor, and that this would concentrate the Orgone energy. Reich called the device an "Orgone accumulator." The earliest boxes constructed of these layered materials were designed for lab animals. He built his first human-sized, five-foot-tall box in December 1940, and set it up in the basement of his house. He began testing the boxes on humans suffering from cancer and schizophrenia - patients were expected to sit inside the Orgone accumulators, completely naked, to facilitate the healing powers of the Orgone energy.

But, since Reich was essentially practicing healing (medicine?) without a license, he came under much scrutiny, and on December 12th, 1941, five days after Pearl Harbor and the day after Germany declared war with the United States, Reich was arrested in his home at 2 a.m. by the FBI and taken to Ellis Island, where he was held for over three weeks. He was later released, but in 1956, after much investigation by the FDA, he would be arrested again. While Reich was in Arizona, one of his associates sent an Orgone accumulator part through the mail to another state, in violation of the injunction, after an FDA inspector posing as a customer requested it. Reich was sentenced to two years' imprisonment. The FDA then descended upon his facility to ensure that all devices and hundreds of copies of his books were destroyed.

On the 23rd of August, six tons of his books,

journals, and papers were burned in the 25th Street public incinerator in New York, the Gansevoort incinerator. The burned material included copies of several of his books, including *The Sexual Revolution, Character Analysis* and *The Mass Psychology of Fascism.* Though these had been published in German before Reich ever discussed Orgone, he had added mention of it to the English editions, so they were caught by the injunction. As with the accumulators, the FDA was supposed only to observe the destruction. An associate of Reich's, wrote: "All the expenses and labor had to be provided by the [Orgone Institute] Press. A huge truck with three to help was hired. I felt like people who, when they are to be executed, are made to dig their own graves first and are then shot and thrown in. We carried box after box of the literature." It has been cited as one of the worst examples of censorship in U.S. history. Wilhelm Reich died in prison before his 2-year sentence was served.

I felt it necessary to include this short (but incomplete) introduction to the work of Wilhelm Reich so that you will understand the wackiness that continues to this day regarding his discovery of the Orgone. I have done much experimentation with many of the devices Reich invented, and I think he was onto something big, but his research was cut short. One of the devices not previously mentioned was his invention of the "Cloudbuster." This device consisted of two rows of 15-foot metal pipes mounted on a mobile platform, connected to cables that were inserted into the running water of a stream. Reich believed that it acted to unblock Orgone energy in the atmosphere, and that it could actually cause rain. Instead of accumulating Orgone energy, this device could work in a reverse fashion. Reich conducted dozens of experiments with the cloudbuster, calling his research "Cosmic Orgone Engineering." During a drought in 1953, two farmers in Maine offered to pay him if he could make it rain to save their blueberry crop. Reich used the cloudbuster on the morning of 6 July, and according to Bangor's Daily News – based on an eyewitness account – rain began to fall that evening. The crop survived, the farmers declared themselves satisfied, and Reich was paid for his services.

The idea of rainmaking has fascinated humans since the beginning of time. Shamans of early indigenous cultures employed many methods including dance to procure rainfall. In the United States, rainmaking was practiced by traveling showmen, and may have reached a peak during the dust bowl drought of the Midwest in the 1930s. Wilhelm Reich's cloudbuster, while considered an unconventional scientific approach, was a new direction in the art of making rain.

Not long after reading of Reich's cloudbuster I decided I had to build one. But, the literature on how to do so was scant, so I was guessing on a lot of the technical details. The old adage, "Keep it simple, stupid" applied here. And it *was* a rather simple affair that could be built from materials easily obtained from the local hardware store. Several long metal pipes, flexible conduit (for grounding in running water), and a wooden frame and platform, were all that was needed to construct the device. Forget the fanciful *Art nouveau* depiction of the cloudbuster in Kate Bush's music video, *Cloudbusting*, this device was crude, but should work just the same. Not long after doing a few experiments with my newly constructed cloudbuster (the results of which were questionable), I remembered a book that was in the archives at Borderland Sciences by a man named Trevor James Constable. The book was, *The Cosmic Pulse of Life*. This book would change everything. Constable, through his research into Rudolf Steiner's concept of the four "Ethers" redeveloped Reich's design of the Cloudbuster. It appeared that Constable had attached Reich's Orgone accumulators of specific geometric proportions to resonantly-designed PVC tubes, which he would then mount on moving vehicles - boats, cars, and later airplanes. According to Constable, this new design negated the need to have the device grounded to a moving water source. I called Trevor to ask for details about his modifications, but he wasn't talking. He did reveal that it could all be figured out by reading certain books - the main one being *The Etheric Formative Forces in Cosmos Earth and Man*, by Guenther Wachsmuth. I was able to obtain a copy of the book and read it four or five times before I set out on my own to try to copy what Constable was doing - much of it derived from the pictures in his books - and the esoteric writings of Wachsmuth. I built around a dozen devices, and then used a friend's farmland to set up some experiments. This was getting complicated. Unfortunately, my friend the farmer liked to gossip, and told several of his farmer friends what I was doing with all that weird apparatus in his field - it soon became the subject of much derision. But, a week into the experiment, the derision turned to amazement. North Dakota was experiencing a drought year, and rain was nowhere in the forecast. But rain it did that week, and all mocking of the experiment subsided. Certainly, the rain could have been mere coincidence, but, to me, this was noble confirmation that the technology worked. I soon thereafter rushed to type up the results of this experiment, and hurried it off to Tom at Borderland Sciences. Tom published the article in the very next issue of the Journal.

What happened next could only be described as exceptionally creepy. Shortly after the article was published, my mother called me to tell me that a man describing himself as "an old friend" of mine showed up at their door, and stated that he needed to get in touch with me. Mom innocently gave him my phone number. Later that same day I got a phone call from a man calling himself "Kirby Thwing." What kind of name is "Kirby Thwing," I immediately thought? I mean, who comes up with a name like that? But, "Kirby" catered to my young ego, and praised my article he'd read in the *Journal of Borderland Research*. We talked for about an hour - he seemed to have a huge amount of knowledge about Reich, Constable, and the whole science of cloudbusting - and he said he needed to visit me in person, and asked where I lived. Hell, I lived just across town from my parents - if he could find them, why couldn't he find me? My kook magnetron started to flash red. I refused to tell him my location, and hung up the phone. I immediately called Tom at Borderlands and told him what had happened. Tom said he thought this "Kirby Thwing" character might be a creepy fucker named Joel Carlinsky, and that he was bad news (It turns out that there was a real "Kirby R. Thwing" who died in a car accident in 2007, and curiously, records show that a "Joel R. Carlinsky" now lives in the same house that Kirby once lived. You can't make this shit up).

The sordid history of this Joel Carlinsky would later

be chronicled in an article written by *Flatland* editor, Jim Martin and *Steamshovel Press* editor Kenn Thomas, entitled, "TOXIC DISINFORMATION: Joel Carlinsky, CSICOP, and Orgonomy." Much of the following data is taken from this article , but one should read it in its entirety:

"....press smear and irrational persecution of orgonomy researchers continues through the agency and covert disruption of Joel Carlinsky of Holyoke, Massachusetts. Carlinsky has written defamatory letters against Reich and those who continue his research, often attempting to pit one against another. He has spied on anyone who expresses a genuine interest in orgonomy by representing himself as a Reich enthusiast, taking advantage of the openness of people who may be unaware of his true intentions. He was arrested, convicted and jailed for burglarizing the Wilhelm Reich Museum in Rangeley, Maine; he has intercepted the mail of medical orgonomists; he conducted his own hate mail campaign against orgonomic researchers with a steady stream of letters to Flatland Magazine, Steamshovel Press and many others; he has even poisoned several by exposing them to radioactive materials."

http://www.orgonelab.org/carlinsky.htm

Where did this clown come from, and why was his sole mission in life to not only discredit the works of Wilhelm Reich, but to attack anyone who was innocently trying to recreate Reich's experiments? And, how was he able to travel the world, showing up on the doorsteps of the authors of articles only a week or two after their publication? The article continues:

"Carlinsky found willing partners in the so-called 'skeptics' clubs, on the Internet, and even in the conspiracy mill. [The] Skeptical Inquirer and its parent organization, the Committee for the Scientific Investigation of Claims of the Paranormal (CSICOP) used Carlinsky as their anti-Reich agent in their quixotic battle against the 'paranormal', recently printing disinformation attempting to connect Reich to the bombing of the Federal Building in Oklahoma City."

This is where the story of Joel Carlinsky gets even nuttier. A few months after I had taken the helm as Editor-in-Chief of the *Borderlands* journal, I got a call from a woman claiming to be from a major TV network. She stated that she had read my article (on weather engineering) in the Journal, and that the network was working on a story about "weather modification," and she asked if I had any photographs of the devices I used that were written about in the article. I said "sure" and sent several of them to the address she had given me. Shit, *Borderlands* was going to be featured on a major network TV show about the amazing experiments we were doing! I had no idea when the show would be aired, and never heard back from her, so I figured the story was dropped and that was that.

Three weeks later, I got a call from one of our Journal subscribers. He asked if I had seen the TV show that a late-night network had done on the Oklahoma City Bombing. Huh? He said the TV show in one part depicted the alleged bombers using "strange weather modification devices" on the farm where they built the fertilizer bomb, and that the two bombing suspects were heavily into the "cloudbusting research" of Wilhelm Reich. The pictures the producers of the show used (claiming to be from the farm of the OKC bombers) were the ones I sent them of my experimental devices I had photographed in North Dakota in 1988. This was my very first introduction to how the major media outlets twisted the truth for the benefit of creating a sensational story, and I was really fucking pissed off. I called the woman from the TV network, expecting she wouldn't answer - she did. I blasted her about her misuse of the photos I sent her - photos I thought were going to be used in a legitimate story on the work we were doing - but were fraudulently used as a prop for their own story on the nutcases of the OKC bombing. I pushed her hard, called her a bitch and a whore of a journalist, and she actually stayed on the line with me - what little conscience she had left, accepted my rather nasty, but warranted assault on her shitty character, until she finally admitted that she got the info - and my phone number - from a informant named, "Kirby Thwing."

Later that year, a guy claiming to be Carlinsky called me at the office. He asked if he would be welcome to drop by Borderlands, "for a friendly visit." Hell, yes, Joel, by all means, drop by for a visit. I arranged the meeting to happen during the middle of the day (2 pm) so that the Borderlands employees would be working in the office - that they may all get a good look at this shithead. He was certainly prompt. At 2 pm, he was at the front door knocking, and I was waiting for him. This guy claiming to be Carlinsky was a very odd looking man - a caricature only an artist could have fabricated - maybe from the imagination of the crippled Lautrec, or the demonic depictions of an H.P. Lovecraft, could this humanoid have been fabricated - from the collective clientele harbored in some dank brothel, or vomiting drunks in a piss-filled alley, or even more likely, from the ranks of some freak-show carnival employment. I acted like it was an honor to meet the infamous bastard, and invited him in. We went upstairs and sat down to have a chat. I grilled him on the OKC thing, and he didn't hesitate to admit he was indeed the informant that gave Borderlands' info to the TV network. I asked about all of the other things he was accused of over the years and he snidely grinned, admitting he was guilty of all charges. If this was the real Carlinsky, he was a narcissist of the first order, and relished any attention that was given to his misdeeds.

After an hour or so of listening to his self-aggrandizing monologue, I politely told him that he wasn't welcome at the office - that we were done talking - and that he needed to leave. His answer to that was to tell me that if he wanted to, he could come back to our office in the wee hours, break in, and make a real mess of things for us. Without hesitation, I opened the top drawer to my desk, pulled out a loaded .44 caliber handgun, cocked the hammer in my lap, and told him, "that would be a real bad idea." He hurriedly left the building, and again, I never heard from nor saw him again.

There's much more on the story of Joel Carlinsky - some of it speculation, most of it true - of his own admission. The thing is, there are droves of Carlinskys out there. He is but a singular classic example of some strange kind of created droidform, employed by some higher agency of control, to misdirect, misinform, and ultimately destroy ingenious and imaginative scientific thought. Unfortunately, in light of the present state of humanity, these dark forces appear to be winning. The propaganda model used by these dark forces has changed little since the inception of its implementation. But, it has been handily thwarted in the not too distant past by a select few who would see through its insidious plan of fear and manipulation.

BORDERLAND SCIENCES ON THE RADIO

My first radio show interview was with an alternative talk show host named Laura Lee in 1991. I had just published an article in the *Journal of Borderland Research* on the history and possible solutions to the "Voynich Manuscript," a medieval tome full of misshapen characters and drawings that had defied decryption since its discovery. I had never done a live radio broadcast interview, and this one was going out to station affiliates across the entire USA. Needless to say, I was quite nervous and was ill-prepared for the barrage of questions the host would slam me with. The "Laura Lee Show" called the Borderlands office to ask if the author of the "Voynich" article would be interested in doing an interview on "cryptography." My youthful confidence answered in the affirmative. That a nationally syndicated radio show wanted me as their guest further bolstered my budding ego, and I was all too willing to offer that inflated ego on the airwaves.

The host of the show, Laura Lee, was a thirty-something blond, a very attractive woman, and she possessed the most incredibly sultry, hypnotizing voice. But, her grounded intelligence and understanding of the alternative sciences made her even more enchanting to my willingness to do the interview. She kickstarted the on-air talk with questions about how I initially discovered the old cryptographic manuscript, and why I felt it necessary to write about its possible solution. The interview was going well, and I was pleased to offer up my inflated egotisms for the thousands of listeners she had collected - I was in total control of the topic being discussed. At some point, I declared that it was necessary to understand certain "algorithms" that were used in the manuscript's decipherment. Laura Lee then asked me,

"Define 'algorithm' for our listeners."

Total silence.

I had no words. I was so stymied by her question - I felt as though everything I had said up to the point of that simple query was purely sanctimonious drivel that I had conjured up from nowhere real, and that I had no business being on the show. And it was true. But, this is exactly how a radio show guest should get his wings, and if he thinks there will be a next show, he'd better humble up and embrace the warranted humiliation of being so ill-prepared.

Of course, I knew what an algorithm was, and should have been able to define it. I simply had a total mental block, and couldn't answer the question. I detected a slight hint of pride in her voice after I delivered that boundless moment of silence. She was kind enough to move along to something else - but only after happily indulging in my puzzled moment of awkward absence to produce an answer to her question. I couldn't have asked for a better initiation into the world of talk radio.

Thankfully, there would many more interviews on the Laura Lee show, in fact, I would become a weekly regular on the radio talk show - her "resident expert" on all things scientifically kooky. One show we did was called, "The Definitive Debate on HAARP." HAARP was a strangely fabricated, but obviously erroneous acronym for "High Frequency Active Auroral Research Project," which was a large plot of land dedicated to a phased-array antenna field on the plains of Gakona, Alaska. Its invention was the brainchild of Dr. Bernard Eastland, and its publically intended purpose was to physically manipulate the ionosphere by heating it to a degree that could manipulate the propagation of electronic communications. Another experiment of this antenna field, that captured the imaginations of several conspiracy researchers, was to modify the weather - as defined in the original Eastlund patents. This government research project initially came under the scrutiny of an Alaskan resident named Nick Begich, and he wrote a very popular book entitled, *Angels Don't Play This HAARP* (with Jeanne Manning, a free energy researcher). In it, he speculated that this farm-field of dipolar antennas could manipulate a portion of the ionosphere such that it could control the minds of everyone on the planet. If that wasn't enough to scare the shit out of people, Begich went on to explain that the device (a small field of radio-frequency antennas) would be used to modify the weather over the entire globe.

Global mind control? Global weather modification? From my understanding and background in radio and radar, it just didn't seem plausible, and I was keen to investigate this project on a deeper level. One of the first things I did was to contact the inventor of HAARP. After a few introductory emails between us, Dr. Eastlund agreed to a phone interview. We dove right in, and Dr. Eastlund told me that his HAARP invention had been far surpassed by the newer technology of Solar Power Satellite based systems which have been designed for a power output of 1000 Megawatts. Let me explain for those nerds with a technical background.

HAARP, does not appear to possess the ability to alter the weather over disparate geographic locations, and has only a minute ability to do anything locally. We can compare this to the natural solar radiation that bombards the earth with great frequency. Radiation from the Sun strikes the atmosphere with a power density of 1370 Watts per meter^2 or 0.137 Watts per cm^2, a value known as the "solar constant." The intensity of HAARP's high frequency signal (2 to 10 megacycles) in the ionosphere will be *less* than 3 microwatts per cm^2 at its full power potential of only 3.6 Megawatts - tens of thousands of times *less* than the Sun's natural electromagnetic radiation reaching the earth. But, these new satellite systems Dr. Eastlund was describing, if indeed they could generate the power proposed, certainly would solve the problem of geolocational weather control. With a fleet of these satellites, it could be conceivable that the weather over any geographic location could be modified at will – provided the system actually works. Borderlands published a book related to this in 1995 called, "Secrets of Cold War Technology: Project HAARP and Beyond," by Gerry Vassilatos. More on "Gerry V." later.

Another author named, Jerry E. Smith, penned a rather kooky book entitled, *HAARP: The Ultimate Weapon of the Conspiracy* (Mind-Control Conspiracy), published by Adventures Unlimited Press. Smith goes far beyond the wild conspiracies of *Angels Don't Play this HAARP,* interpreting his intuitional and uneducated understanding of the technology as some doomsday machine that has been implemented in many locations all over the planet.

After much informative correspondence with Dr. Eastlund, and also based on my hands-on knowledge of radio technology - that made the conspiracy hypothesis seem completely absurd, I proposed an on-air debate with these two authors to Laura Lee. She thought it was a grand idea, and suggested the debate to both Begich and Smith. They accepted, and the show was on.

I had spoken to both Begich and Smith before this show, and had argued my points technically - neither of them had anything in the way of substantiation for their theories. The live debate on the Laura Lee Show started out with a very basic account of the topic at hand, and then we dug in. The very first thing I found out was that neither of the two I was engaging on this topic of HAARP had *ever corresponded with or spoken to Dr. Eastlund.* If you are going to write a book about someone else's invention, I would think the first thing to do is contact the inventor. Nope, they'd never talked to him. I then asked them both,

"So, guys, do either of you have a background in radio technology?"

They both replied, "No."

Both authors repeatedly invoked the name of Nikola Tesla throughout the pages of their books - that HAARP was actually the product of Tesla's inventions, but neither had any idea what Nikola Tesla had actually done. While HAARP was designed to heat the ionosphere, Tesla emphatically stated in his famous June, 1919 article in the *Electrical Experimenter* entitled, “Famous Scientific Illusions,”

“...there is no Heaviside layer (ionosphere), or, if it exists, it is of no effect.”

Written some years after his Wardencliff Project, Tesla explains the inefficiency and limitations of the Hertzian method of electromagnetic propagation (through the air) with great clarity — and that his system of wireless which was vastly more effective, utilised the ground itself for propagation. Tesla states, "Properly constructed, my system [of wireless] is safe against static and other interference, and the amount of energy which may be transmitted is billions of times greater than with the Hertzian which has none of these virtues..." He explains in particular, with several analogs in diagrammatic representation, his single-wire-without-return system — the heart of Tesla's radio and wireless power systems.

We know the “Heaviside layer,” now called the ionosphere, is the radio-reflective layer that exists within the upper atmosphere, and is presumed responsible for radio propagation over great distances. Since the HAARP Project's exclusive purpose relies on the existence and function of the ionosphere, and has nothing to do with the grounded circuits of Tesla - it is clearly *not* Tesla technology. It seems many “kook out” at the mere mention of Mr. Nikola Tesla.

You can imagine how the debate went after all of this. Why would these authors writing about the HAARP Project make such gross mistakes — mistakes which destroy the credibility of any other statements they have made concerning the technology? One might suggest that because these people are journalists and movie makers, and not radio technicians or electrical engineers, that we are supposed to forgive their glaring errors. But a good journalist is expected to present the truth as clearly and as disinterested as is possible. And, one would think a good (ethical) journalist would not overlook the facts, in favor of a sensational presentation, unless the books were merely intended to be works of fiction. Most of the associations these days concerning Tesla and sinister military technologies come from later statements made by Tesla - long after his direct involvement in experiment with wireless. Speculations and predictions by Tesla such as microwaves, TV, beam ray technologies (the Tesla Death Ray), cosmic-ray motors, interplanetary communications, and wave-interference devices (the "Tesla Howitzer" and the "Tesla Shield") have been given authority over the actual physical experiments by Tesla. Most of these conjectures were revealed by Tesla in his aging years at "birthday press conferences" as he was shut out by the daily media sometime after 1920. But, these are the statements that so capture the attentions of readers and viewers, and are the ones which the so-called "underground media" gleefully disseminate. Gloom and doom is the status quo here, and while the truth may be stranger than fiction, it is apparently quite boring.

What was curious to me, was that even after I published the revelations Dr. Eastlund offered - the fact that HAARP had been made obsolete by space-based satellite technology of much greater power - none of the HAARP conspiracy authors investigated or wrote about any of it. Here was a disclosure of huge proportions, especially for the conspiracy writing crowd, and none of them jumped on it, even though my publication of Dr. Eastlund's admissions (with his permission) had gone "viral" on the internet (in an age where the term "going viral" had not yet been invented). But, one science writer we regularly published in our Journal, did jump on it. That was Gerry Vassilatos.

GERRY VASSILATOS: THE BRILLIANT SCIENCE TEACHER OF STATEN ISLAND

Gerry Vassilatos had already begun working with Borderland Sciences shortly before my return in early 1991. His three-part series of videos - "The Ray of Discovery" had become legendary in the alternative sciences. Some viewed the videos as the product of pure genius, others couldn't make any sense of it. Each part in the series was over two hours long, consisting of Gerry's heavily echoed narration to a background of still images, transitioned with fades and glides reminiscent of a Ken Burns documentary. This would have been fine but for Gerry's treatment of the images - a true kaleidoscope of colors and saturated pictures one would imagine seeing at the peak of an LSD trip. Some loved it, but most absolutely hated it - and told us so. The information presented by Gerry in his narration was unprecedented - it would seem as though Gerry knew the secrets to all of the scientific mysteries, and his eerie delivery only emboldened the notion. Each series began with Gerry in silhouetted relief reciting the following mantra - astonishingly reminiscent of an Orson Wells oratory:

"There is a power loose in the earth...it is ancient and forever new. No one may resist it. It is the very Ray which gives Discovery. No one will stop, or manipulate it. The Ray of Discovery will change this world. Hello again, and thank you for tuning in. In our series of telecasts, we will chronicle some of the most astounding discoveries made in our time. You and I together will

rediscover technology which has been kept hidden by seditious privateers, who in their greed and malice, have vainly attempted to manipulate and alter the natural movement of discovery. In our series, you will be shown astounding discoveries which have been suppressed for some fifty years. Look to your viewing screens now, as we unfold this drama...together."

The series of videos is now available on YouTube - and Vimeo, and I suggest you go watch them - one has to experience *The Ray of Discovery* both visually and audibly in order to assimilate the greatness of its performance and content.

See: https://vimeo.com/19577546

I began having phone conversations with Gerry sometime in late 1991. Gerry was an encyclopedia of knowledge in the arcane sciences - he seemed to know quite a bit about everything from Alchemy to Zoroastrianism with a huge smattering of Reich and Tesla thrown in between. Gerry was also a musician, and created all of the spooky tracks for the *Ray of Discovery* videos on his Mellotron, which he had acquired from none other than Jan Hammer. The Mellotron is an electro-mechanical, polyphonic tape-replay keyboard originally developed and built in Birmingham, England, in 1963. It was really the first sampling instrument, as each key on the keyboard would play back pre-recorded sounds through a tape recorder head affixed to an individual key. Jan Hammer had used it to great effect with the Mahavishnu Orchestra in the 1970s, and after Gerry acquired it from him, it was used to transform

Gerry's musical ideas into etherial soundtracks for his videos, and another series of musical albums that Gerry designated by number with the preceding title, "Muudo."

Our conversations quickly escalated from music to the lost sciences and in no time we were collaborating on several research projects. We would continue this collaboration for many years until Gerry's ultimate disappearance from the alternative science scene in late 1999.

Upon hearing both my album *Mundus Subterraneus* and Gerry's *Muudo* albums, a fellow research associate and artist named Duncan Laurie invited us to his "Dragonline Studio" (duncanlaurie.com) in Newport, Rhode Island. This would be the only in-person meeting I would ever have with Gerry Vassilatos.

Duncan is a sculptor and glass artist who had in the past created several large works for the corporate elite. His focus in later years would turn toward the melding of science and art - with a specific emphasis on the science of radionics. There are far too many erroneous definitions of radionics available to anyone who would do a search on the topic - I'll try to explain what it really is, here. Radionics is the science of defining the qualities of "living energy." So how do we define "living energy?" The life-force inherent in all of us is hardly quantifiable, but through the use of specific devices that attempt to quantify this elusive life-force, the science of radionics was born. Radionic history really dates back to the age of the ancient megalithic architects, the primordial craft of the shaman, and back to a time when this life-force was heralded as an intrinsic part of our very being. But the term "Radionics" came from late

19th and early 20th century researchers - a time when radio was born.

Radio had captured the imaginations of nearly everyone on the planet - it was a magical new technology that brought the news and story-time into people's living rooms every evening. Radios of the era were handcrafted works of art, Art Nouveau and Art Deco masterpieces in wood, metal and glass, each fitted with a chassis filled with components as mysterious as the sound of the radio announcer's voice emanating from the tabletop box. The knobs that turned dials affixed to the front of these instruments became the tactile progenitors of a world-view unprecedented in previous history. Tom Lewis', *Empire of the Air* is a must read, or a watch of Ken Burns' video documentary version of the same, is essential to all radio enthusiasts. The magic of radio is fully captured in this work.

With this advent of radio came a host of new visionaries of a radio-like science to be used in a new way - the medical arts. The pioneer in this new field of "medical radionics" was a man named Albert Abrams, and he would be the first to develop a device called, "The Dynomizer" which looked much like the conventional radio of the times, and Abrams claimed it could diagnose any known disease from a single drop of blood or even from the medical subject's handwriting. He performed diagnoses on dried blood samples sent to him on pieces of paper in envelopes through the mail. Abrams even claimed he could conduct medical practice over the telephone with his machines, and that he could determine personality characteristics with the device. He gave his devices names such as the "Oscilloclast" and "Radioclast" and they were constructed of very distinctive radio-like components - two double-wound electromagnets and a pendulum mechanism that could deliver approximately 200 interruptions a minute. This interrupter supplied power to an air core transformer, which was essentially a Tesla coil that was short circuited. The output of this coil was connected to 3 parallel switched resistor banks. Each resistor in these banks was 1000 ohms, a value Abrams felt was in resonance with organisms, and the output of each bank was connected to a single post which was then electrically connected to the patient. Thus, the patient was in series with the circuit and the resistors.

Other radionic practitioners would follow. Ruth Drown created an instrument, again resembling the radio receivers of the day, she called the "Homo-Vibra Ray." This device was a complete departure from Abrams' instruments in that no commercial electricity was ever introduced into its circuit. From her book on the subject, *Drown Radio-Vision and Homo-Vibra Ray Instruments and Their Use*, she stated:

"It has been difficult for the average physician to recognize the great vital life force which flows through everything in the universe, giving it its own rate of vibration. When that rate of vibration changes, the substance recognized as a definite thing ceases to be that thing and becomes something else without ceasing to be a substance. Were commercial electricity used with this instrument, we would defeat our own purpose to the degree that we would waste the electric energy, for it would not find its own rate of vibration in the human body, and we would also obscure the flow of human energy to such a degree that the human patient would

receive only a portion of the treatment that he otherwise would receive were he to treat with his own energy unobscured."

Trevor James Constable was also a huge fan of Ruth Drown, and had this to say of her work:

"The key to understanding the work and instruments of Dr. Ruth Drown lies in a firm grasp of the two fundamental energy manifestations connected with the human body. These are, first, the electrostatic energy of the physical-mineral tissues, and second, the vital energy which passes into and through those tissues enabling them to hold their form. Electrostatic energy remains with the physical body after death. It is the departure of vital energy from the physical body that constitutes death."

The evolution of this healing science called Radionics, derived from the basic advent of radio technology, was interesting not only for its intellectual ingenuity, but for the inventor's quest to reveal the wholly qualitative aspects of our life-force from quantitatively-derived measurements. Both Abrams and Drown would face hostile opposition and legal prosecution for their work. They were both labelled as therapeutical villains - pseudoscience kooks who were grifting their patients out of house and home, all from a medical establishment moving ever towards an industry of false pharmaceutical panaceas brought on by the greed associated with the massive income they would procure. Nevermind any positive results Abrams and Drown would achieve.

Borderland Sciences had republished much of Abrams' and Drown's research on Radionics, and Duncan Laurie, having read these works, showed a keen interest in it. Duncan understands that Radionics is much more than just the circuits of old electrical devices, and he has been able to transcend the gap between this science and the various art mediums he works with.

When Gerry and I arrived at Duncan's studio in Newport, we were immediately astonished by the amount of glass that was built into the construction. Floor-to-ceiling windows adorned every side and level of the three-story studio, and the floor of each tier of the building was made of glass block - you could see through the very ground you were walking onto the story below. The ground floor was consumed with racks of electronic equipment, and Gerry and I focused our attentions there. These racks were filled with military WWII era radio transmitters and receivers, all in working condition. Since both Gerry and I were into strange analog electronic musical instruments, we set about creating a "theremin" using two of the giant shortwave receivers. The theremin, originally known as the ætherphone, is an early electronic instrument that requires no contact from the musician playing it. It is named after the Westernized name (Léon Theremin) of its Russian inventor, Lev Termin, who patented the device in 1928. It consists outwardly of two metal antennas, one used to change the volume of the instrument, and one for changing the pitch. The musician simply places their hands in proximity to these antennas, and by varying the distance from them, will effect change in either amplitude or frequency of the sound emitted from the device.

Gerry and I understood the superheterodyning technology of radio inventor Edwin Armstrong and we immediately set to work on creating a theremin-like device from two of the shortwave military receivers. It worked quite brilliantly, and could be played as an electronic instrument (I had previously used this very setup on my album, *Mundus Subterraneus*).

It was a short-lived but very good meeting with Gerry, and it cemented our future research relationship. We collaborated on several projects over the next 7 years - many having to do with ground radio applications, and the bio-electric nature of plants. In some cases, 80% of the journal would be written by Gerry and myself - under several pseudonyms such as "Franklin Ellsworth Clarke," "Charles Allon," and "Marion Delia Worthing." Gerry was the most prolific author I'd ever worked with, and I published his seminal works, *Secrets of Cold War Technology*, *Lost Science*, the seven-volume set of *The Vril Compendium*, and the last, *Declassified Patents of the Cold War and SDI*, which would spawn further writings and videos, the remainder of which would eventually end up never being published. These last writings were not for the lighthearted uneducated conspiracy theorist, no, the material contained in them could only be assimilated by the most ardent of serious researchers. Gerry and I discussed the content at length - nearly every day for months, and because of the sensitive content - we argued over its publication. I maintained (however mistaken, I later realized) that because of the feeble sales we had with the previous works, these new documents had little potential for commercial success. I also questioned the "declassified" aspect of the material. None of this info had ever been presented in any fashion before, and seemed quite risky. It was around this time that Gerry called me one day and said he was finished with Borderlands and finished with all of the research he had done to date. No real explanation was given. Since he was my go-to writer and the best researcher and author I'd ever encountered, I naturally pleaded with him not to go away. After we hung up the phone that day, not I, nor any others heard from him again.

This event, coupled with a few more impossible fucking tragedies would spell out the coming future of alternative science publishing, and would end with my eventual departure from the helm of Borderland Sciences Research Foundation.

What could possibly make a man who had dedicated his entire life and soul to the research of lost science in a single day simply disappear himself from it all together? My reasons for leaving have, over some time, become clear to me, and I have reconciled that decision. I have very uncomplicated and altogether transparent reasons for jumping ship at that time, but Gerry never really offered up a straightforward explanation for his action. We can only speculate, and and the following may offer an answer.

A very large portion of the issues one encounters on this path of disseminating alternative scientific information is the onslaught of vitriolic dissent one must endure from not only the academic intelligentsia, but from the very heartbeat within our own ranks. We can accept that we will be derided and dismissed by those who do not question anything other than what they have been spoon-fed by credentialled academia, but when we are viciously attacked by those among our own brethren…

There is far more dissent and ridicule erupting from within the alt-sci community than there is from without. It is the most destructive element in any society - to be cleft in twain by its own living body of seeming commonality, and generally erupts from the vices of vanity and jealousy, and in some cases, just plain insanity. These are the meetings with remarkable kooks, whether they be from within or without, they are still pure kooks. Now, we shall nuzzle up to the very bosom of kookdom, and that suckling nest from which they emerge - the unwholesome cradle from whose diminution and confinement they have wholly embraced to become the embodiment of illness, and an anathema to the furtherance of all that is right with the vessel of alternative science.

Y2K, ART BELL, AND RAMS

Back in 1998, I wrote an article called, "Y2k and New Industry of Hysteria." One of my colleagues rightfully proclaimed that the 'Industry of Hysteria' was nothing "new," and she was correct in thinking so. So correct in fact that her disparagement of my use of the word "new" in the title of the article, forced my proposal to her. We were quickly married, and shortly thereafter, quickly divorced. It is of little consequence regarding the forthcoming story.

In 1983, I was introduced to the personal computer. I had just started my second year in the Marine Corps, and one day after payday while wandering around the Post Exchange (PX) on base (the Post Exchanges sell consumer goods and services to authorized military personnel), I came across a display featuring the new "TI-99/4A" personal computer. It was priced around $350, and I grabbed a box off of the top of the display, and just bought it. When I got the computer home, I proceeded to dive right in and start programming. My subject for the first program I would create? The Tarot! Yes, the very first computer program I wrote was a Tarot card reading application. My grandmother had introduced me to the Tarot when I was a teen, so I a had a pretty good understanding of what this divinatory oracle was about.

My knowledge of how to create a program with the graphics necessary to make it an interactive experience was non-existent, but after reading the documentation, I was able to portray a rudimentary representation of what is referred to as the "Celtic Cross" reading. That was actually the hard part. The easy part was creating the data, or the "meanings" of the cards to be selected at random from the usage of the built-in pseudorandom number generator (PRNG) that I programmed into the Tarot application. After 36 hours of continuous coding, my first program was finished. It was a very poor portrayal of the esoteric fortune-telling card game, but it worked as advertised.

Later in the 80s, I would try my hand at creating new algorithms for graphic fractal generation, and I went on to create some novel data encryption programs. At first, I wrote simple programs for substitution ciphers, and then they got more complex as I returned to using the pseudorandom number generator in the algorithms. But pseudorandomness was not good enough for me - being pseudorandom was not true randomness, and keys generated with PRNG could conceivably be cracked by present technology. I knew of a more secure method of encryption. It's called the One-time Pad. In cryptography, the One-time Pad (OTP) is an encryption technique that cannot be cracked, but requires the use of a one-time pre-shared key the same size, or longer, as the message being sent. In this technique, a plaintext is paired with a random secret key (also referred to as a One-time Pad). Then, each bit or character of the plaintext is encrypted by combining it with the corresponding bit or character from the pad using modular addition. If the key is truly random, is at least as long as the plaintext, is never reused in whole or in part, and is kept completely secret, then the resulting ciphertext will be impossible to decrypt or break. I would then go on to write the first OTP encryption program for the DOS operating system.

In 1989, I got into creating computer bulletin board systems (BBS). A bulletin board system or BBS was a dial-up connected computer server running software that allowed users to connect to the system using a terminal program. Once logged in, the user could perform functions such as uploading and downloading software and data, reading news and bulletins, and exchanging messages with other users through email, public message boards, and sometimes via direct chatting. I ran several BBSes from 1989 to 1994 - the content of them would include all manor of science and technology topics, and several were all about computer programming and hacking. The last BBS I managed was the "Borderlands BBS." You can find evidence of it here: http://bbsmates.com/viewbbs.aspx?id=117054

This brief history of my involvement in the personal computer movement would later be the catalyst for writing about Y2k.

The Y2k article I wrote in '98 focused on all the hype surrounding the "The Year 2000," and how computers and everything else with some kind of digital control system would cease to function. I first published the article in *The Journal of Borderland Research* and the response to it was less than gratifying. Computer experts came out of their digital caves in droves to disparage the dispatch I had meaningfully crafted to calm the public fear - fear that was being inflamed by writers, journalists, and talk radio hosts who had little understanding of basic computer functions and hardware. These disparagements were as easily shrugged off as the derision received on many occasions regarding much of the material we published. But there was something else. Other publishers, looking for an alternative viewpoint on Y2k, were asking permission to republish the article in their own magazines. And, because I wanted to get my viewpoint out there, I gave these publications carte blanche to do so. The article was republished in no less than 12 different magazines - most of which would eventually publish a retraction of the article - stating they were misinformed by the writer. Their published retractions would appear in editions of their magazines long before the bell tolled midnight on January first, the year 2000. Apparently, they had received so many negative letters about my article, and many from so-called "credentialed experts" that they all felt it necessary to print a retraction, in most cases stating they were misled by what I wrote, and that my information on Y2k was completely wrong. It turns out it was spot on, but very few listened or believed it.

It was around this time that I discovered other late-night talk radio programs - specifically "Coast to Coast" hosted by Art Bell, and "Sightings" hosted by Jeff Rense. These talk shows and their hosts truly embraced the worlds of alternative science, and the guests they interviewed were a direct reflection of late-night talk radio kookiness. Guests such as Richard Hoagland (the "Face on Mars" discoverer), David Oates (Reverse Speech pioneer), Gary North (Y2k Gloom and Doomer), and Ed Dames (Remote Viewing) were regulars on the show, and it was a great source of late-night entertainment. But, something about these shows really started to bug me. Here we were nearing the end of the millennium, and the advertising on commercial breaks was all about survival. Ads for wind-up radios and a year's supply of food went along perfectly with the doom-and-gloom ideology the guests were offering in their lyrical mantras over the AM airwaves.

If you were a listener in the late '90s it was a time of wild conspiracy theories and fabricated prophecies offered to listeners with very few solutions save buying something that they advertised. It was enough of a catalyst to engender a willful response from my distaste for the subject matter, and respond I did.

This is where the people who have anticipated my recount of the so-called "Talk Radio Feud" will want to read carefully. There won't necessarily be any new revelations or disparaging comments about anyone involved. Except maybe those few who are dead - but we say nothing wicked about the dead, do we? Most everyone from that era of wild talk shows is still alive and will probably read this.

Late in 1998, around the time I had written the Y2k article, I was contacted by a man named Robert A.M. Stephens. Robert was extremely complimentary about the recent articles I had written, and was keenly interested in talking about - talk radio. Robert professed to be a former Navy SEAL, and a contractor for NASA - working as a "failure control specialist" or some such title like that - and an artist. Robert was a painter and utilised techniques that any Bob Ross aficionado might use on the canvas, but he certainly had passion for his subjects, and much of his work reflected that.

Around the same time, I fell in with a group called the "Mass Depopulation Trio." MDT later became what was a loose group of hacker-types, and had taken over the "alt-fan-art-bell" IRC (Internet Relay Chat) group. This internet chat room consisted of fans of Art Bell and a group of characters who absolutely hated him. After looking at what people were saying in the chat room, I rather quickly fell into the crowd that hated Art Bell. And then, well, I was hooked. A cadre of fools, maybe, but it was a cadre of friendship.

Stephens was convincing as hell, so much that he commissioned me to create a website for him, and I accepted. If Robert was anything he was convincing. I was a former Marine and his initial dialogue about his experience in the Navy SEALs was quite convincing ... military speak at its finest. Except it was all bullshit. Someone online had raised the question about his military background, and I got right on it.

Let's go back to the first time I'd talked to Robert (his nickname was "RAMS"). After several weeks of chatting online with this character called "RAMS", he gave me his phone number and told me to give him a call. A few hours later, I dialed the number. A young girl answered the phone.

I asked if Robert was there, and the girl said he was in the shower. I explained who I was and why I was calling, and she proceeded to talk my ear off about Robert for the next 30 minutes. She told me story upon story about Robert, and that most recently he had been arrested for driving around in his favorite Jeep shooting at inanimate objects with a sawed-off shotgun. According to her, Robert showed up in court for the offense with his dog, and told the court he was representing himself. When asked questions about his charges, he deferred to the dog as an eyewitness for questioning, as his dog was present during his arrest. She told me the judge was so disgusted with the entire ordeal that he threw the case out. She told of another time when Robert was in a local diner and 3 guys decided it would be ok to ridicule Robert for dressing like a cowboy. One of them decided to take Robert's Stetson from his head and play hat-catch with the other two. She said Robert got up and stabbed the man with a fork. The girl told me - a complete stranger - these and more tales all the while Robert was apparently in the shower, and I never did actually get to talk to him that night.

The next day, Robert called me and said, "all of that shit is true." He talked about his years as a Navy SEAL, and all the work he had done for NASA. As I said before, it was all very convincing. But, after a few people had heard these claims by Robert on his first appearance on the Art Bell show, they began to question the validity of said claims.

The first thing I did was contact a group of former Navy SEALs dedicated to outing fakers. I submitted Robert's full name to them, and they returned in email that in fact, they had no record of a Robert Stephens or a Robert A.M. Stephens ever having attended or completed BUD/S training. BUD/S is a nice acronym for "Basic Underwater Demolition/SEAL training - the Navy SEAL's basic training course, and anyone claiming to be or have been a Navy SEAL will have attended this basic training. RAMS was nowhere to be found in their database in any form of his name. What follows is a transcript of the email thread of my investigation:

Robert A.M. Stephens Exposed as a Fraud

I have recently done a fairly thorough investigation into the claims made by Robert A.M. Stephens. Borderland Sciences' association with Mr. Stephens began after he emailed me about information I had posted to my website concerning the December 7th Landing (a supposed UFO landing - more misleading info). Some of the information Mr. Stephens was presenting seemed very interesting (although unverifiable) and I asked him if he would like us to put up a webpage for him.

After the December 30th, 1998 guest appearance by Mr. Stephens on the Art Bell show, many (including myself) started to question Mr. Stephens credibility.

The following is a report of my investigation on Mr. Robert A.M. Stephens.

Initially, I checked out Robert A.M. Stephens (after his first email to me) by doing a typical search on the internet for his name. I came up with the following references:

http://www.nasa.gov/hqpao/library.html/arts/samples.html

http://digester.et.tudelft.nl/IMp66-15.html

http://digester.et.tudelft.nl/IM66-15.html

http://www.d21.k12.il.us/~lists/fb1744.html

http://mirror.apple.com/mirrors/00_links.for.anarchie.dir/Info-Mac.Archive/per/im/infomacv15/infomacv15-051-100/infomacv15-066.txt

I have to admit that my investigation was not very thorough. Mr. Stephens gave me a phone number to a

"Larry Mauk", his contact at KSC, that turned out to be a fax machine. I sent a fax asking for confirmation of Mr. Stephens, but received no reply.

After the Art Bell show appearance, I began to do a more thorough investigation into his background. I emailed some of the contacts he said he had at NASA. One of them, declared a "buddy" by Mr. Stephens, wrote back saying, "I have never met or talked to Robert Stephens to my knowledge, and I have no idea if he is a Space Shuttle contractor as stated in his e-mail." He then went on to say, "I was simply responding to his e-mail in the spirit of public mail to NASA, and am a bit chagrined that he has termed me his "buddy" and that he posted my e-mail to him to the Web without asking me if he could do so." This NASA spokesperson asked that I do not forward or post his mail without permission. I am awaiting his permission.

I then went after the SEAL Team-3 claim. There are many websites on the internet hosted by former Navy SEALs which enjoy exposing those who make such false claims. Here is a list:

Fake SEALs and Phoney Frogs
Phonies and Scumbags
Fake SEALs
Wannabe Veterans
Mike's US Navy SEAL Homepage

There may be more. I then emailed ALL of them asking for confirmation whether Robert A.M. Stephens had or had not been a Navy SEAL. Here are a couple responses so far:

From: "VeriSEAL" <xxxxx@ix.netcom.com>
Organization: Phony SEAL Hall of Shame
To: Michael Theroux <director@borderlands.com>
Date: Mon, 4 Jan 1999 20:49:52 +0000
Subject: Re: More info from ROBERT STEPHENS
Reply-to: xxxxx@ix.netcom.com
Priority: urgent
X-mailer: Pegasus Mail for Win32 (v3.01b)

Michael,

*All of your e-mail info on Stephens has been forwarded to about a dozen *REAL* former SEALs. What you have in the form of Stephens is a 100% bonafide FRAUD.*

Look for incoming messages from assorted retired Viet-era SEALs who will likely be more than interested in Stephens. By the way, the only two Teams in existence during WW-Vietnam were 1 and 2.

The owner of the Hall of Shame at the Teams Homepage (http://The-South.Com/TheTeams) is retired Senior Chief Kent Dillingham, one of the plankowners of ST-3 in 1983. That means he was one of the original founding SEALs for that Team.

Ciao, Dana

I then Emailed Robert Stephens with this (his reply follows):

Date: Mon, 04 Jan 1999 08:44:57 +0000
From: "Robert A. M. Stephens" <sti3818@montana.com>
To: Michael Theroux <director@borderlands.com>,
Paul Dore <pdore@web-computing.demon.co.uk>
Subject: Re: Hi Robert-Hi Michael

Michael Theroux wrote:

>Hi, excuse me for doing my homework, but this is what I have >come up with:

>West Coast Teams
>SEAL Team 1 - South East Asia
>SEAL Team 3 - Middle East

>SEAL Team 5 - South Korea

>East Coast Teams
>SEAL Team 2 - Northern Europe
>SEAL Team 4 - South America
>SEAL Team 8 - Africa

>SEAL Team 6 - Anywhere

>According to my sources, Team 3 wasn't formed until after the >Nam thing. So, there seems to be a conundrum here...

>Is this correct?

>just checkin' you out.

Stephen's response:

No problem, if they care. No conundrum.

Teams by formation save for 6 and 8, are pretty much open for deployment by locale. Team 3 formation was in 1970 or thereabouts. In the deploy for Courier Services and other interdiction duty in non Viet Nam battle zone areas such as Laos, Cambodia, Thailand and northern Burma and up on the Chinese border proper, several small elements from Teams 2, 3 and 4 were used or detached. The Courier Services is where I entered into service under 'last surviving son status' (see the selective service link and the Military archive link). My draft number in 1970 was 311 out of 366. The women and children would have gone before me and the public thought for Viet Nam [sic] at that time was the same view it has now for child molestation. Viet Nam was winding down too military wise.

The statement above for region location is something new to me. I did not know and have not been aware of the designation you show above such as 'South Korea' in relation to Team number. However, this may be something more recent.

The resolve where Teams would be located did not come finally until 1975 and is still subject for relocation for center locale. In 1982 it was rumored they would all be combined on the south coast. Just rumors. 6 and the new 8 are units I know nothing about since they are really under the actual true umbrella 'ops' types operations and I do not know their details or what they do. Team 2 practiced for ship interdiction and was the true resolvers for the Achilles Laurel ship hijacking though the media credited the Marines. However, they do everything else too not reserved for that above.

The Pol Pot connection with the Khmer Rouge with support of the Chinese communist served as the last and most severe potential faction for continued upset in the southeast asia region as Viet Nam [sic] was winding down for America. I never saw or was in Viet Nam ever.

Robert Stephens

I then forwarded this to the VeriSEAL contact and this was his reply:

From: "VeriSEAL" <xxxxx@ix.netcom.com>
Organization: Phony SEAL Hall of Shame
To: Michael Theroux <director@borderlands.com>
Date: Mon, 4 Jan 1999 21:06:01 +0000
Subject: Re: More info from ROBERT STEPHENS

Michael,

For starters, you indicated his resume claims attachment to ST-3 from '70-'74. ST-3 didn't exist at that time. Secondly, he is nowhere to be found in the SEAL personnel records archives.

Lastly, as demonstrated by his message which you were kind enough to forward, he doesn't have the slightest idea what he's talking about.

His name will eventually be prominently posted in the Hall of Shame.

Ciao,
Dana

And finally, I spoke with Mr. J. H. "Hoot" Andrews, USN, Ret. Retired from SEAL Team TWO.

He reiterated what I had already heard, and verified that Robert A.M. Stephens is a fake.

I called Mr. Stephens after this and asked him some questions about this. He said he had never heard of Senior Chief Kent Dillingham, one of the plankowners of ST-3 in 1983 (see above) and the very SEAL team of which Stephens was supposedly attached. I asked him about the VeriSEAL's claims, and why if he felt these people were wrong, did he not take issue with them?

He said he did not care.

I told Mr. Stephens that I would be forced to take down his website because of his fraudulent statements, and he replied that that was fine and that I should expose him as a fraud, fake, pathological liar, etc. I agreed. He did say he would be forwarding a letter from Larry Mauk confirming his contracting work with NASA (nothing ever showed up).

I admit that I should have investigated him more thoroughly from the start - as Richard Hoagland said on Art Bell's show, “On the internet, nobody knows you're a dog!” This is really quite true, but I must say that a good investigation will reveal a lot.

While Mr. Stephens appears to have been deluded, it by no means makes all of his points and statements, invalid. But, how are we to believe anything he has said in light of this information? Does the “NORAD/Fence” database exist? Does the “Blind Transom” DoD tracking software exist? What about his alien implant information? Probably not.

Now, we are left with the question, “What was Robert A.M. Stephens' motivation for all of this?” Was it for his “15 minutes” (actually an hour) of fame? Or, was it some sinister cover up plot devised by whomever he may be really working for?
Perhaps we'll never know.

January, 04 1998
Michael Theroux
Director
Borderland Sciences Research Foundation - Since 1945

Shortly after sending my investigation of RAMS to Art Bell, he called me and asked me to be a guest on his radio show. He did want to briefly discuss the RAMS info, but had heard that I had written a Y2K article and wanted to discuss that. For the next five hours we discussed all manner of Borderland Sciences subjects.

Stephens would later tell me he had been recruited by certain agencies including NASA to go after Richard Hoagland and Art Bell in an operation secretly named, "Troxxes," to discredit the information they were disseminating on the Coast to Coast radio show, and that all of the fake information he provided was simply a ruse to open up the op. It would appear that I had been had again. Robert A. M. Stephens apparently died in 2013, and I guess we'll never know the real story. Oh, I never heard from him again.

MASS DEPOPULATION TRIO

The Mass Depopulation Trio (MDT) organically grew from the roots of the disinfotainment.com website, which was a blog and so much more. MDT started putting together audio mashups of talk radio show host dialogs and mixing them with certain sound effects and snippets of songs. Some of the music was actually composed for these so-called "spams." MDT would consist of three pseudonymous characters: "MickeyX", "Johnny Pate", and "Dr. HD Slow, all of whom had a devilish ability on the internet to make a mockery of, and virtually destroy any and all resident kooks who were always threatening to call the FBI. And, I'm sure many of these kooks did so.

While MDT was a trio, they all seemed to work independently and became involved in several shenanigans that would later become legend. Of the greatest achievements of MDT, "Mel's Hole" would win hands down.

Mel's Hole is, according to an urban legend, an allegedly "bottomless pit" near Ellensburg, Washington. Claims about it were first made on the radio show *Coast to Coast AM* by a guest calling himself "Mel Waters." Later investigation revealed no such person was listed as residing in that area and no credible evidence that the hole exists. From the Wikipedia site on Mel's Hole:

The legend of the mythical bottomless hole started on February 21, 1997, when a man identified as Mel Waters appeared as a call-in guest on Coast to Coast

AM with Art Bell. Waters claimed that he formerly owned rural property nine miles west of Ellensburg in Kittitas County that contained a mysterious hole. According to Bell's interviews with Waters, the hole had infinite depth and the ability to restore dead animals to life. Waters claimed to have measured the hole's depth to be more than 15 miles (24 kilometers) by using fishing line and a weight. According to Waters, the hole's magical properties prompted US "federal agents" to seize the land and fund his relocation to Australia.

Waters made guest appearances on Bell's show in 1997, 2000, and 2002. Rebroadcasts of those appearances have helped create what's been described as a "modern, rural myth". The exact location of the hole was unspecified, yet several people claimed to have seen it, such as Gerald R. Osborne, who used the ceremonial name Red Elk, who described himself as an "intertribal medicine man...half-breed Native American / white", and who told reporters in 2012 he visited the hole many times since 1961 and claimed the US government maintained a top secret base there where "alien activity" occurs. But in 2002, Osborne was unable to find the hole on an expedition of 30 people he was leading.

Local news reporters who investigated the claims found no public records of anyone named Mel Waters ever residing in, or owning property in Kittitas County. According to State Department of Natural Resources geologist Jack Powell, the hole does not exist and is geologically impossible. A hole of the depth claimed "would collapse into itself under the tremendous pressure and heat from the surrounding strata," said

Powell. Powell said an ordinary old mine shaft on private property was probably the inspiration for the stories, and commented that Mel's Hole had established itself as a legend "based on no evidence at all".

Mel's hole was a complete fabrication of a couple of the members of MDT, with a certain member acting out the part of "Mel" as a guest on the Coast to Coast radio show. Several more "hoaxes" would be fabricated and presented on the show. One of the more notable ones was The EQ Pegasi Hoax.

In late October, 1998, the BBC in London ran a story that a possible SETI (Search for Extraterrestrial Intelligence) signal from the nearby star called "EQ Pegasi" had been found by an UK amateur astronomer using his company's 10 meter satellite tracking dish. Art Bell and his frequent guest on *Coast to Coast AM*, Richard Hoagland, seized on the opportunity to tell wild tales about this incident.

In the latter years of my tenure as Director of Borderland Sciences Research Foundation, I found myself in the odd role of debunker of bullshit science. While I was fully invested in the qualitative, esoteric side of science, I still knew total bullshit when I observed it. And, I wasn't about to ignore it.

This so-called amateur astronomer had created a website detailing the information on exactly how he had detected this alien signal. On the website were two screenshots of the software depiction of the signal - at two different times of the day. Almost immediately I noticed a huge problem with his graphics. In each of the two graphic representations of his signal detection, the lines of the signals were slightly different. Cool. But the problem with these graphics was in the background noise. The background noise of space radio is always truly random - no two images will ever be alike. His background noise was exactly the same in both graphics. Busted! I sent my findings to both the SETI Institute and to Art Bell. The SETI Institute responded that I was spot on and that it was a complete hoax. Art Bell didn't respond at all and continued to propagate the hoax night after night with more fabricated stories. I was pissed. In the following days I hammered Art Bell's website with further evidence I had that the event was a hoax. No response. On November 3rd, I discussed the claims of the bogus signal on the Laura Lee Show. And, on November 5th, myself, Nathan 'Chip" Cohen (astronomer and engineer with experience in SETI) and Seth Shostak (then Director of the SETI Institute) talked about the claims on the Laura Lee show. We had completely debunked the EQ Pegasi signal detection, but none of this deterred Art Bell and his guest from continuing the insane charade on his nightly talk show.

While re-researching the EQ Pegasi event for this book, I was blown away at how completely fucking stupid the entire spectacle was, and what a complete waste of time it was.

To engage oneself in debunking an obvious hoax is oftentimes an endeavor in complete futility. But, in this case, it was only the beginning in a long list of calling out the charlatans, and in the end, they conceded defeat. The kookfactor was at its zenith in the late '90s and MDT was a grand catalyst for setting things right.

Please have a look online - it's still archived for any who'd love to waste some time - or have a keen interest in its anthropological significance.

MDT virtually disbanded shortly after Y2K never materialized. Their work was done, and so was the credibility of late-night talk radio kookdom.

FUCKING MONTANA

Soon after all the fanfare and the "nothing happened" of Y2k, I left the publishing company. I was seduced away mostly by a friend I had worked with in the past on some shitty bank security software (I wrote the encryption routines, and he wrote the database part). He had landed a job with a large financial institution, and called me up one day to ask if I wanted to move to Minneapolis to work with him. I really didn't want to leave Borderlands, or Northern California, but when he threw out some salary figures, I had to think about it. The kook factor I was immersed in had drilled a jaundice-colored foot up my ass for the last year and a half, and I was quite sick of it.

So I popped. I had a week to resolve things with my publishing company, pack up what I could, and bail NorCal for MLPS. I filled my newly acquired Crown Vic's giant trunk with everything I could fit in it, briefly said goodbye to my soon-to-be wicked girlfriend, and drove almost straight through from Eureka, California to Minneapolis, Minnesota. It was a long fucking trip. I drove up through Yellowstone, and then I hit Montana. If you've ever driven through Montana from end-to-end, you will never want to do it again. Going east, you take the 15 up to Butte, the 90 to Billings, Miles City, and after what seemed like ten fucking days, I was in North Dakota. No matter how you approach Montana from end to end, it is a purely hellish driving experience. The only consolation is the raw beauty the "big-sky" state offers up to the hearts of the more esoteric. The mountains and the truly big sky meet with one another in a gorgeous embrace at dawn, and again they meet in solemn repose at dusk. At several points during the journey through the state, you have to stop and gaze on into its wonderment. It is a state like no other in the union of all the United States. But driving through it still sucks, and if you make it all the way through, you have accomplished something few have ever encountered. Then, North Dakota. I grew up in North Dakota - on the east side of the state - in Grand Forks. I can think of no place more barren a wasteland than Grand Forks. Except maybe sometimes what's in my head driving through Montana. There is a certain contagion that one can become infected with as regards the mental instability of those who one associates with, and from the utter loneliness of those you ignore. It becomes readily apparent when reviewing what one has recently written. The grammar goes out the fucking window, and the words make no sense upon having re-read them - they are but the jumbled masses of inarticulate thoughts scribbled and scrawled on the pages in paragraphs so wide and so narrow, poor Euclid would be right to proclaim asininity! Oh, but they are the antithesis of Euclid's musings - more they are the endless "planes-at-infinity" of non-Euclidean geometries. These are some of the thoughts a kook scientist has when traversing the vastness of Montana, only to be greeted by the infertile holocaust of that which is the landscape of North Dakota. A lovely reward for such a daring adventurer's conquest.

Only as I passed the "Welcome to Minnesota" sign did I regain composure and ultimately, my sanity. Passing through Fargo/Moorhead (I lived there when I was 5) is grand, because for the last 835 miles you've been travelling nearly straight due-east. Fargo/Moorhead signals the end of that with I-94's detour south, and the 5 hr. homestretch to Minneapolis.

I arrived at my final destination at 8:30PM on a Sunday evening - the following Monday morning I would be thrust into the new life I had chosen for myself - the life of corporate minion, or as I later would call myself, a "corposlob." What had I done? It had been my thought that being the director and publisher of Borderland Sciences, with its rich history, would be my last job I would ever have. But, there was no turning back now, and the culture shock I would soon experience, was an unprecedented assault on everything I believed in.

ABOUT THE AUTHOR

Michael Theroux is an Author, Artist, Musician, Researcher, and Martial Arts Instructor. From 1985 to the year 2000, Michael was the Managing Editor, and later the Editor-in-Chief at Borderland Sciences Research Foundation. He authored hundreds of articles in the Foundation's quarterly magazine, *The Journal of Borderland Research*, and is the author of several books including, "Rhythmic Formative Forces of Music," “Biological Communications” and "Y2k Hysteria." Michael was a guest on Art Bell's "Coast to Coast AM" radio programme, and a regular guest on "The Laura Lee Show." Since the year 2000, Michael concentrated his talent working with several of the top Fortune 500 companies (Primevest, ING, IBM, Plexus Corp.) to address their IT security needs. Michael Theroux is currently an independent consultant focusing on computer security. Michael’s interests include, but are not limited to, cryptography, malware forensics, microkernels, physics, music, cartography, navigation, horology, climatology, and polar exploration. He currently lives in Wisconsin.